The
Marshall Cavendish
International
WILDLIFE
ENCYCLOPEDIA

VOLUME 14
MAN–MOS

MARSHALL CAVENDISH
NEW YORK · LONDON · TORONTO · SYDNEY

Introduction by: Professor D.L. Aller
General Editors: Dr Maurice Burton and Robert Burton
Consultant Editor: Mark Lambert
Production: Brenda Glover
Design: Eric Rose
Editorial Director: Nicolas Wright

Revised Edition Published 1990

Published by Marshall Cavendish Corporation
147 West Merrick Road
Freeport, Long Island
N.Y. 11520

Managing Editor: Mark Dartford BA
Editor: Nigel Rodgers MA
Production: Robert Paulley BA
Design: Edward Pitcher

Printed and bound in Italy by LEGO Spa Vicenza

Library of Congress Cataloging-in-Publication Data

Marshall Cavendish International wildlife encyclopedia/general
    editors, Maurice Burton and Robert Burton.
        p. cm.
        ''Portions of this work have also been published as The
International wildlife encyclopedia, Encyclopedia of animal life and
Funk & Wagnalls wildlife encyclopedia.''
        Includes index.
        Contents: v. 14. MAN-MOS.
        ISBN 0-86307-734-X (set).
        ISBN 0-86307-000-0 (v. 14).
        1. Zoology–Collected works. I. Burton, Maurice, 1898-
II. Burton, Robert, 1941-   . III. Title: International wildlife
encyclopedia.
QL3.M35 1988
591'.03'21–dc 19

Volume 14

Mantis

The name Mantis is derived from a Greek word meaning 'prophet' or 'soothsayer' and refers (as also does the epithet 'praying') to the habitual attitude of the insect—standing motionless on its four hindlegs with the forelegs raised as if in prayer as it is waiting for unwary insects to stray within reach. 'Preying mantis' would be a more suitable name. The forelegs are spined and the joint called the tibia can be snapped back against the femur, rather as the blade of a penknife snaps into its handle, to form a pair of grasping organs which seize and hold any unfortunate victims.

Mantises, or praying mantises as they are often called, feed mainly on other insects, and are found mostly in tropical or subtropical countries. Most of the smallest are about an inch long. They have narrow, leathery forewings and large fan-shaped hindwings, which are folded beneath the forewings when not in use. Most mantises can fly, but they do not readily take to flight and seldom go far.

About 1 800 species are known, the most familiar species being the European mantis **Mantis religiosa**, which lives in the Mediterranean region and has been introduced into eastern North America.

Hidden terror

Most mantises spend their time sitting still among foliage, or on the bark of trees, waiting for insects to stray within reach of a lightning-quick snatch of their spined forelegs. Nearly all are shaped and coloured to blend with their surroundings. Many are green or brown, matching the living or dead leaves among which they sit, but some have more elaborate camouflage which serves two purposes. First, because they do not pursue their prey but wait for it to stray within reach, they need to stay hidden. Secondly, their grasping forelegs, although formidable to other insects, are usually useless against birds and lizards, and since mantises are slow-moving, they must be

Like an unknown monster from outer space the mantis cradles a day-flying moth in its spined forelegs and delicately and neatly eats its live victim.

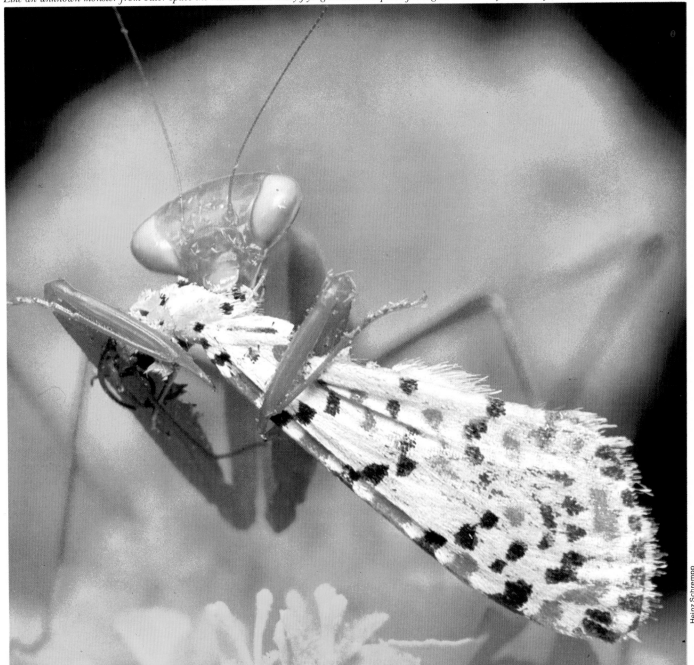

Heinz Schrempp

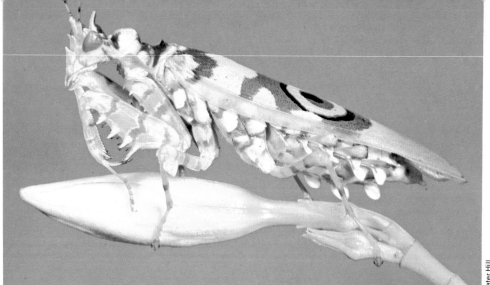

Soothsayer — but beware

Its typical pose may be one of prayer but don't be taken in by its reverent appearance for the mantis is really a voracious creature ready to grasp out at any potential food with a quick snap of its spined jack-knife claws.

Peaceful meditation — two mantids **Mantis religiosa** (below) pose in their characteristic resting position with head and prothorax held high and forelegs folded in an attitude of prayer. But this serene picture is deceptive for as soon as any suitable victim appears within reach these mantids will lash out viciously to seize it.

Head gone, now for the rest! It is well known that the female eats the male after or during mating, he is just another meal to satisfy her apparently never-ending appetite (above right). She grasps him with her penknife-like forelegs and eats on without a qualm.

African mantis, **Sphodromantis** sp., with its egg sac or ootheca attached to a twig (above far right). The eggs are laid in tough spongy capsules for protection against predators.

An exotic African mantis **Pseudocreobotra** shown to full advantage (above left and left). The magnificent eye-like markings on the wings serve a definite purpose in threat display. When frightened the mantis will spread its wings so any enemy comes face to face with another 'face' that is both unexpected and very menacing.

Ready for blast-off! Not quite, for this rather bizarre position is one of threat display. Sitting up on its hindlegs and with forelegs stretched out the mantis **Hierodula** stands over its fallen victim (right).

The drawings (below far right) give a comparison of the penknife-like forelegs of the mantis (top), mantis-shrimp (centre) and mantis fly (bottom). Three invertebrates dissimilar in every other respect except in the way they catch their prey.

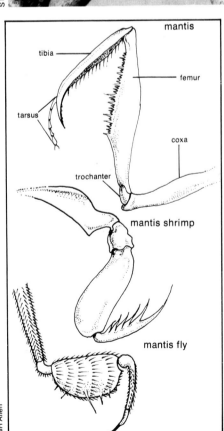

mantis

tibia

femur

tarsus

coxa

trochanter

mantis shrimp

mantis fly

Anthony Bannister: NHPA

◁ *As the female* **Sphodromantis gastrica** *lays her eggs she gives out a liquid which she stirs into a froth by movements of her body. The plastic-like substance then hardens enclosing the eggs. Each capsule, which is usually attached to a twig, contains about 80–100 eggs. One female may make as many as 20 such oothecae in her lifetime.*

▽◁ *Section through an ootheca of* **Sphodromantis** *sp. showing the egg chambers. The spongy texture of the capsule is seen clearly here but it nevertheless gives good protection against hungry birds.*

▽ *Young mantids emerge from the ootheca. They may moult up to 12 times before becoming adult. Their wings, tiny at first, grow larger with each moult.*

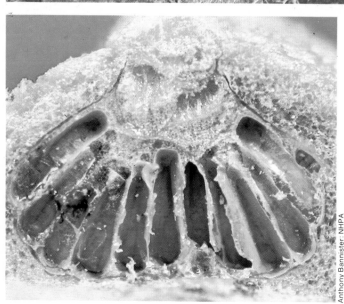

Anthony Bannister: NHPA

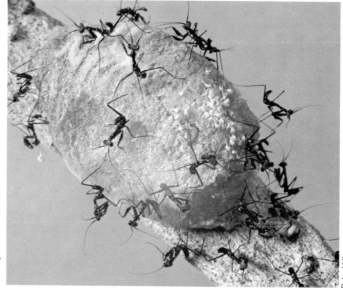

Peter Hill

concealed to avoid being caught and eaten.

Mantises never take plant food. They seize their insect victims in their spined forelegs and eat them alive, neatly and delicately. Some of the largest species occasionally catch and eat small birds and lizards in the same way.

Lose your head—and your inhibitions

To a female mantis a male is no more than just another piece of food. He must, therefore, be careful in his approach if he wishes to mate, rather than be the next meal. On seeing a ripe female, the male, justifiably enough, freezes, then starts to creep up on her with movements almost too slow for the eye to follow—sometimes taking an hour or more to move 1 ft. Once within range, he makes a short hop and clasps the female, to mate. If the pair is disturbed or the female sees her suitor, she will eat him, starting by biting off his head. As he loses his head, so he loses his inhibitions, because mantis copulation is controlled by a nerve centre in the head which inhibits mating until a female is clasped. If this nerve is removed (by an experimentor, or by a female mantis) all control is lost, and the body continues to copulate. The female, therefore, has much to gain from attacking and eating males; she ensures both fertilisation of her eggs and nourishment for

her developing ovaries.

Eggs in a bag

The female lays 80–100 eggs at a time in tough, spongy capsules which she attaches to twigs, and she may produce 20 capsules in her lifetime. While laying her eggs she gives out a liquid which she stirs into a froth by movements of her body. The eggs become enclosed in this while it is still plastic, then it quickly hardens and dries.

The young mantises hatch together and at first hang from the egg capsule by silken threads which they give out from the hind end of the abdomen. After their first moult they can no longer make silk. They grow by gradual stages, moulting up to 12 times before becoming adult. The wings, tiny at first, grow with each succeeding moult.

The egg capsules are a protection against insectivorous animals and birds, but they are no protection against parasitic wasps of the ichneumon type, which are probably the most serious enemies of mantises.

Fatal flowers

Some mantises are even more deceptive, taking on the appearance of flowers and so luring insects such as bees and butterflies within reach. The orchid mantis of Malaysia and Indonesia, in its young or subadult stage, is coloured pink and the thigh joints

of the four hind legs are widely expanded so they look like petals, while the pink body resembles the centre of the flower. When the mantis reaches the adult stage, however, its body becomes white and elongated as in a normal mantis. It still has the expanded, petal-like legs but its resemblance to a flower is largely lost. The African 'devil's flower' has expansions on the thorax and the forelegs which are white and red. It hangs down from a leaf or twig, and catches any flies or butterflies attracted to it.

When they are frightened, many mantises will suddenly adopt a menacing posture, rearing up and throwing their forelegs wide apart. One African species *Pseudocreobotra wahlbergi* improves on this display by spreading its wings, on which there are a pair of eye-like markings, so the enemy is suddenly confronted with a menacing 'face'.

class	**Insecta**
order	**Dictyoptera**
suborder	**Mantodea**
family	**Mantidae**
genera & species	***Hymenopus coronatus*** *orchid mantis*
	Mantis religiosa *European mantis* ***Idolum diabolicum*** *African devil's flower, others*

PH Ward

Mantis fly

Like so many insects loosely called flies, the mantis fly is not related to the true, two-winged flies, or Diptera, but belongs to the order Neuroptera, to which the alder flies and lacewings also belong. Mantis flies are fairly small, rarely as much as an inch long. These rare insects live only in the tropics and subtropics and even there one is lucky to see one. Nevertheless, they have an extraordinarily peculiar and specialised life history. In one respect at least mantis flies resemble the praying mantis (p. 1541), although they are not related; they catch insects in their penknife-like front legs. The wings of the mantis fly have a fine network of veins—a characteristic of the alder flies and the lacewing flies too. Also, the forewing is normal, not a leathery covering for the hindwing, as in the mantis, and the hindwing is not folded and pleated.

Food and lodging provided

The larvae of the rare mantis flies are parasites on wolf-spiders and wasps or other insects, while the adults live among foliage, preying on small insects. The southern European species *Mantispa styriaca* is one of the few species that has been studied in some detail. Its eggs are rose-red in colour and are laid on long stalks of silk like those of a lacewing, looking like tiny pins in a pincushion. The larvae that hatch from them are known as 'campodeiform', which means 'like a *Campodea*', one of the bristletails (p 421). In fact the larvae do look like small silverfish or tiny earwigs, as do the larvae of other Neuroptera. They go into hibernation immediately after hatching, becoming active again the following spring, and starting to look for the egg cocoons of wolf spiders *Lycosa*. The female wolf spider spins a white silken cocoon for her eggs and carries it with her wherever she goes, fastened to the tip of her abdomen. A single *Mantispa* larva enters each cocoon and feeds on the young spiders as soon as they hatch, becoming very fat and swollen by the time it has devoured them all. The mother spider notices nothing unusual and continues to carry the egg-sac about with her, unaware of the havoc going on.

△ ***Mantispa interrupta*** *seizing a fly with its penknife-like front legs (4 × natural size).*
▽ *Larvae, about 1·5 mm long, emerge from their eggs only to parasitise spiders' eggs.*

Anthony Bannister: NHPA

Multiple metamorphosis

The mantis fly larva then moults its skin and re-appears in a caterpillar-like or 'cruciform' shape, with a relatively tiny head and small legs, wholly unlike the original hatchling. Still within the spider's egg sac it spins a cocoon among the dried remains of its victims and pupates. Just as the larva exists in two distinct and dissimilar stages, so does the pupa which, as well as an inactive stage inside the cocoon, has a second, active stage outside it.

Insects in which the larva or pupa shows more than one form in the course of development are said to undergo 'hypermetamorphosis', metamorphosis beyond the normal number of stages. A Brazilian mantis fly *Symphasis varia* has a similar life history, but its larvae live as parasites in wasp nests.

Variations on a theme

The development of a front leg in which the end joint snaps back, onto the one behind, like a penknife blade, has taken place independently in the mantis flies and the mantises. It is found in at least two other kinds of arthropods, a fly and a crustacean. The New Zealand wingless fly *Apterodromia evansi* has forelegs in which two of the joints, the tibia and the femur, are developed almost exactly as in a mantis. This fly belongs to a family of predators, the Empidae. Certain marine crustaceans known as mantis shrimps also have spined grasping claws of this kind. Taken together these make up a truly impressive case of what is known as convergent evolution, in which a similar feature is developed independently for a particular purpose by several unrelated animals.

class	**Insecta**
order	**Neuroptera**
family	**Mantispidae**

Anthony Bannister: NHPA

Mantis shrimp

The most striking feature of this lobster-like shrimp is its pair of mantis-like claws used for seizing prey, and from which it derives its name. The last joint of each limb closes like the blade of a jack-knife on the preceding joint. In some species this last joint is armed with sharp spines, while in others the 'blade' has a smooth cutting edge fitting into a groove in the preceding joint. These pugnacious animals, of which there are less than 200 species, may be from 1½ in. long to over a foot so they must be handled with caution.

The jack-knife claws are second in a series of limbs of various kinds on the thorax. They are preceded by a small slender pair and are followed by three much smaller pairs of clawed limbs used in digging. Three unclawed, slender pairs of limbs follow these, and on the abdomen are five pairs of broad, flat swimmerets with filamentous gills. The swimmerets of each pair have small hooks which link them in the midline of the body. The final pair of limbs, as in a lobster, forms the sides of a large tail fan. In some species, the abdomen is flattened and widens towards the hind end making the body appear oddly proportioned. More-over, the abdomen looks longer than it really is, because the dorsal shield, or carapace, does not cover the last four segments of the thorax as it does in the lobster. The large abdomen houses some of the internal organs more commonly found in the thorax in related crustaceans. The front part of the head has two hinged parts, one carrying the eyes which are on stalks, and the other the antennae. The carapace has a hinged beak (rostrum).

Mantis shrimps are usually brilliantly coloured, green or brown; although some are blue or red, mottled or with alternating bands of colour, or with peacock's eyes of azure blue and gold on the tail.

They are also known as split-thumbs, squillas, prawn killers, shrimp mammies or nurse shrimps. Mantis shrimps are marine, particularly common in tropical seas, and are rarely found at depths of more than 1 500 ft.

Marine guillotines

Mantis shrimps seize their food with a lightning movement of their claws, hold it fast with the spines, when these are present, and tear it apart. One sharp-clawed specimen in captivity was seen to cut a shrimp in two with a single stroke, and the pieces fell to the bottom together, as if still united. Many mantis shrimps rarely leave their burrows but lie in wait at the entrance for small fish, shrimps, crabs, worms, molluscs, sea anemones, and other mantis shrimps, which they eat. Others venture out more often to pursue their prey, especially at night, propelling themselves through the water with the swimmerets on the abdomen.

Prickly subjects

Mantis shrimps live mainly in shallow water or on the shore, making deep vertical or sloping burrows in sand or mud or concealing themselves in holes or crevices in rock or coral. One species *Gonodactylus guerinii*, which lives in coral, plugs the entrance to its hole with its specially modified and spiny tail fan. The spines not only give protection in the obvious way but also make the visible portions of the mantis shrimp look like a sea urchin attached to the surface of the coral. Other species with a similar tail fan defend themselves by lashing out with the abdomen.

Devoted mothers

When mating the male places his sperms in special pockets on the underside of the female's thorax near the pair of openings on the sixth segment where the eggs will be laid. These are small, $\frac{1}{40}$ in. diameter, and are glued together by a cement produced by glands near these openings. They stick together in a mass which the female carries on the three pairs of small clawed limbs and constantly turns over and cleans. The swimmerets are not used to hold the

Gonodactylus sp., claws at the ready, waits to snatch out at any small fish passing by.

eggs, as they are in many Crustacea. In some species the female lays the eggs in her burrow where she guards them.

The larvae swim or drift about, moulting at intervals before eventually becoming bottom-dwelling adults. Some of the larvae in tropical seas are over 2 in. long with a striking transparent, glassy appearance. In some places they are so numerous that the sea looks like a thick soup.

Gladiatorial fights

Two of the common names of mantis shrimps are prawn killers and nurse shrimps. These two names express totally different ideas, and while the female mantis shrimp is a devoted mother the fact remains that a fair amount of cannibalism and fighting takes place within the species. When fighting they lash out with tails and legs but seldom do each other serious injury. A fight is always preceded by a threat display in which the two shrimps spread their legs and claws, exposing certain white spots and silver streaks. It is something of a puzzle why a mantis shrimp should be so heavily armoured and armed and yet live in a burrow that it leaves only occasionally to feed. At night it even plugs the mouth of the burrow with pieces of coral and anything else that is handy. It seems, however, that the armour of the spines is not so much a defence against predators as a means of defending a territory. Sometimes the lawful owner of a burrow is ejected, but only after a lengthy fight, and usually when the occupant is a little weak, for example when it is moulting.

phylum	**Arthropoda**
class	**Crustacea**
subclass	**Malacostraca**
order	**Stomatopoda**
family	**Squillidae**
genus & species	*Squilla desmaresti* *S. mantis*

Marabou

With its large, heavy bill and pink, almost naked vulture-like head and neck, the marabou is one of the uglier storks. It stands 4 ft high with a wingspan of over 8 ft. The back, wings and tail are dark grey, the legs are mainly grey, and the underparts are white. A most unattractive pink, fleshy pouch dangles from the throat. This is apparently part of the respiratory system but has now become a secondary sexual character, rather like the wattles of domestic chickens.

The marabou is found in Africa from Sudan and Ethiopia to Zambia, being especially common in East Africa. Closely related to it are the adjutant and lesser adjutant storks of Asia from India to Borneo. The name adjutant has also been given to the marabou, possibly because of the stork's pompous strutting gait.

▷ One marabou sees a joke but the pelican and other marabou can only see the serious side.
▷▽ Reflections on the matter—marabous stand majestic and motionless in the water.
▽ Balding head, moulting down and ugly pouch hanging from its throat—an unfortunate bird!

Toni Angermayer

Just like vultures

Marabous are in many ways like vultures. The naked head and neck appear to be the same adaptation for carrion-eating as in vultures and condors (p. 644), as feathers would become matted with blood while feeding on a large carcase. Marabous and vultures often mix around carcases of big game, but marabous haunt human settlements where there are carcases of domestic stock or smaller refuse; they flock around slaughterhouses and rubbish-dumps. On the shores of the large East African lakes they gather by fishing villages to feed on offal left over from cleaning the catch. Because they perform the services of a public health department, marabous and adjutants are allowed to roam in cities and in some places are protected by law.

Scavenging brings marabous into contact with other carrion feeders such as vultures and hyaenas. It is quite a common sight to see all three together at a carcase. The marabous dominate, stealing choice pieces of flesh which they have ripped from the carcase. Large pieces of bone may be swallowed. The marabou is also the chief predator of flamingos and will, at times, take the whole crop of chicks.

When seeking carcases of wild game marabous may have to fly long distances. Like many other large heavy birds, they do this by soaring in thermals, rising currents of warm air. This is the method used by vultures and other birds of prey which need to conserve energy. They can glide effortlessly over long distances, but this method restricts takeoff until the sun is quite high and the air has warmed up.

Breeding when animals die

Unlike most tropical birds, which breed in the wet season when there are abundant plants or insects on which to feed their young, marabous breed in the dry season. The advantage of this is that during the dry season their food is concentrated in a few places. The lakes and rivers are low so fish and frogs are easily caught and the game concentrates around waterholes and is killed by predators. In settled areas, more cattle are killed in the dry season because of the shortage of food and ploughing makes it easy for the marabous to catch mice. Marabous nest within easy soaring and gliding distance of such sources, assuring an easy food supply for the chicks.

At the beginning of the breeding season marabous gather in their nesting trees. The males get very aggressive and their throat pouches are inflated. Females trying to join the males are driven off at first, but after many tries are allowed to approach their prospective mates. Once a pair has formed the male sets about collecting sticks, which the female weaves into a nest about

Des Bartlett: Photo Res

Simon Trevor: Photo Res

3 ft across, with a lining of twigs and green leaves. The nest is never left unattended as other marabous would not hesitate to steal sticks from it.

Nest building takes a week or more and sticks are added throughout the season. When the main construction is finished 2–5 chalky white eggs are laid. Both parents incubate them until they hatch 30 days later. The newly-hatched chicks are almost naked and their parents brood them or shade them from the sun for a fortnight. After that they are still covered in bad weather. Until the chicks are quite large one parent always stands guard to prevent neighbouring marabous from removing the nest from under the chick.

Marabou chicks are fed by both their parents. Food is usually dropped on the nest for the chicks to pick up and water is regurgitated into their mouths. Very few birds bring water to their young but it is a common habit in storks and is apparently due to the chicks' habit of excreting onto their legs, which helps cool their bodies by evaporation of the fluid.

When they are about 15 weeks old, the chicks begin to fly from branch to branch of the nest tree, flying from the tree a week or so later.

A balanced diet

Young marabous are fed many frogs, tadpoles, fish and mice as well as scraps of meat and offal, and this is one reason why marabous nest within easy reach of stretches of open water. Sometimes flying termites and locusts are brought to the nest.

The value of these extra items in the diet was shown by MP Kahl who, in the course of his studies on the marabou, attempted to hand-raise three chicks. For the first few days they fed on raw meat, then they began to refuse it. but readily accepted fish and frogs. The chicks grew well, but when 7 weeks old the fish trap used to catch their food was stolen. They again refused to eat raw meat alone and rapidly grew weaker. Their bones became extremely brittle, showing that whole animals are needed to supply calcium for their bones. Raw meat does not contain much calcium, and the bones of big game are too large for the chicks, so small animals, complete with skeletons, supply their calcium.

class	**Aves**
order	**Ciconiiformes**
family	**Ciconiidae**
genus & species	***Leptoptilos crumeniferus*** *marabou*
	L. dubius *adjutant stork*
	L. javanicus *lesser adjutant*

Vultures and marabous feast themselves on a dead giraffe. Scavengers alike, it is quite common to see these birds and hyaenas all feeding together.

Marine iguana

The marine iguana is unique in its way of life, being the only truly marine lizard. It is found only in the Galapagos Islands, some 600 miles west of Ecuador. Because of its exceptional home, it is of great interest, but physically it is not so exciting. The accounts of early visitors to the Galapagos testify to the marine iguana's ugly appearance. One account describes them as having the most hideous appearance imaginable, and the same author, a captain of the Royal Navy, says that 'so disgusting is their appearance that no one on board could be prevailed on, to, take them as food'. Marine iguanas grow up to 4 ft long. They have blunt snouts, heavy bodies, clumsy-looking legs with long toes and a crest that runs from the neck to the tail. The tail is flattened sideways and is used for swimming. Most marine iguanas are black or very dark grey, but on Hood Island at the south of the Galapagos Archipelago their bodies are mottled with black, orange and red and their front legs and crests are green.

'It's a hideous looking creature of a dirty black colour, stupid and sluggish in its movements. The usual length of a fullgrown one is about a yard, some even 4ft long.'
Voyage of HMS Beagle, *Charles Darwin 1890.*

Lizard heaps

Outside the breeding season, when they are not feeding at sea, marine iguanas gather in tight bunches, sometimes even piling on top of each other. They lie on the lava fields that are prominent but unpleasant features of the Galapagos. In the heat of the day they seek shelter under boulders, in crevices or in the shade of mangroves. At the beginning of the breeding season, the males establish small territories, so small that one iguana may be on top of a boulder while another lies at the foot. Fights occasionally break out but disputes are generally settled by displays. A male marine iguana threatens an intruder by raising itself on stiff legs and bobbing its head with mouth agape, showing a red lining. If this does not deter an intruder, the owner of the territory advances and a butting match takes place. The two push with their bony heads until one gives way and retreats.

While marine iguanas are basking, large red crabs will walk over them, pausing every now and then to pull at the iguanas' skin. The lizards do not resent this pulling and pinching and with good reason, because the crabs are removing ticks from their skin. Darwin's finches (p 751) perform the same service.

Diving for a living

As the tide goes down the marine iguanas take to the water and eat the algae exposed on the reefs and shores. They cling to the rocks with their sharp claws, so as not to be dislodged by the surf, and slowly work their way over the rocks tearing strands of algae by gripping them in the sides of their mouths and twisting to wrench them off. At intervals they pause to swallow and rest. Some marine iguanas swim out beyond the surf and dive to feed on the seabed. They have been recorded as feeding at depths of 35 ft but usually they stay at about 15 ft. The length of each dive is about 15—20 minutes but they can stay under for much longer. When Darwin visited the Galapagos in HMS *Beagle* he noted that a sailor tried to drown one by sinking it with a heavy weight. An hour later it was drawn back to the surface and found to be quite active.

Marine iguanas normally eat nothing but marine algae. Unusual exceptions are the marine iguanas that haunt the home of Carl Angermeyer. He has trained them to come at his whistle to be fed on raw goat meat, rice and oatmeal.

Harems round the rocks

When the males have formed their territories, the females join them. They are free to move from one territory to another but the males soon gather harems of females around them and mating takes place without interference from other males. Courtship is simple: a male walks up behind a female, bobbing his head, then grabs her by the neck and clasps her with his legs.

When the males leave their territories, the females gather at the nesting beaches. There is competition for nest sites and fighting breaks out. Each female digs a 2ft tunnel in the sand, scraping with all four feet. Sometimes they are trapped and killed when the roof falls in or when a neighbour scrapes sand into the hole.

Only 2 or 3 white eggs, $3\frac{1}{4}$ by $1\frac{3}{4}$ in., are

△ *Like a lichen encrusted monument — the marine iguana presents his best side to the camera and shows off his metallic colours. In profile his snout is seen to be blunt and the clumsiness of his legs and heavy body is apparent. But against the blue sky the red and green mottling of this otherwise rather grotesque reptile is given full due.*
▽ *Diving for food. The marine iguana is the only modern lizard that uses the sea as a source of food. It is herbivorous, feeding exclusively on seaweeds.*

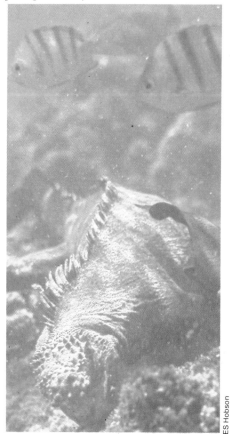

laid. Then the female iguana fills up and camouflages the tunnel. When the eggs hatch in about 110 days, 9in. iguanas emerge.

Apart from man the main enemies of full grown marine iguanas are sharks, but the iguanas usually stay inshore where sharks are not likely to venture. Young iguanas are caught by herons, gulls and Galapagos hawks, as well as introduced cats.

It's warmer out

While he was on the Galapagos, Darwin found that it was impossible to drive marine iguanas into the sea. They would rather let themselves be caught than pushed in and if thrown into the sea they would hurriedly make for the edge and clamber out. This is a surprising habit for an aquatic animal as most animals that habitually swim, such as turtles and seals, make for the safety of the sea when frightened. Darwin assumed that the marine iguana behaved in this strange way because it had no natural enemies on land but that the sharks were waiting for it in the sea. If this were so, it would mean that marine iguanas would have to be pretty hungry before setting out to feed. Recently another explanation has been put forward. While basking, marine iguanas regulate their body temperatures to within a range of 35—37°C/95—99°F (see Bearded lizard p 311 for a discussion on temperature regulation in lizards). The sea temperature around the Galapagos is 10°C/50°F less, so the marine iguanas are reluctant to escape into the sea, as this makes them too cool.

class	**Reptilia**
order	**Squamata**
suborder	**Sauria**
family	**Iguanidae**
genus & species	*Amblyrhynchus cristatus*

Marine toad

Sugar cane has been spread, from an unknown native home in the Far East, to warm countries round the world, including tropical America. With it went the grey cane beetle, the larvae of which live on the roots and can destroy the canes. In tropical America lived the marine toad which feeds on cane beetles. In an effort to control the beetle the marine toad has been taken round the world, wherever sugar is grown. On the whole it has failed to control the beetle, but for the toad itself this is a success story.

The marine toad does not normally live in the sea. It was presumably so named because one was first seen by the sea. Its habits are similar to those of more familiar toads, like the common European toad, and at first sight its main claim to fame is its size; it is sometimes called the giant toad. Usually it is 5 in. long and weighs ¾ lb, but females, which are larger than males, may be up to 10 in. long and weigh 3 lb. It is mottled yellow and brown in colour and has numerous dark reddish-brown warts on its body.

Its native home is from southeastern Texas, through Mexico and Central America to South America as far south as Patagonia. It has been introduced into Florida, many of the islands of the Caribbean, Bermuda, Hawaii, New Guinea and Australia, and to a few other places where sugar cane is grown.

Nocturnal toad

During hot weather the marine toad, also known as the cane toad in Australia, remains under cover of vegetation or burrows in the ground, coming out at night, or in wet or cool weather, to feed. It will take anything that moves which is small enough to swallow, especially insects and beetles, and also smaller toads and frogs, small snakes and even mice.

Rapid growth

One of the more unusual features of this toad is that some males change sex later in life and become females. Mating, however, is normal, by amplexus in water. Each female lays several batches of eggs, up to a total of 35 000 in a year. The young toads, after completing the tadpole stage, grow rapidly and are said to reach a length of 5 in. after a year.

Toxic toad

When it is attacked, the marine toad is said to be the most poisonous of all toads; its poison may cause closed eyes and a swollen face, with other unpleasant symptoms such as nausea and vomiting. In extreme cases it may result in death as the poison acts like digitalis and slows the heart beat, so leading to heart failure. The poison is a whitish fluid exuded in small doses from a patch of glands on each side of the head. The toad blows itself up when disturbed, as so many toads do. Although it has no warning colours, in its native home it has few, if any, enemies. In countries where it has been introduced the native predators have not the instinct to leave it alone, largely because it has no warning colours, and those that try to eat it either suffer from its poison or are suffocated when, having quickly gulped down the toad alive and whole, it blows itself up and blocks the throat or the gullet of the unfortunate predator.

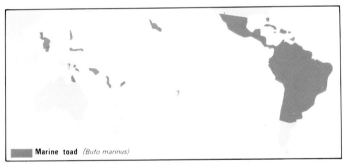

Marine toad (Bufo marinus)

Haughty stare from a disgruntled marine toad. Contrary to its name this toad does not live in the sea but stays hidden under vegetation or burrows in the ground on hot days, coming out at night to feed. With a large appetite to satisfy the marine toad eats all sorts of insects and anything that moves and is small enough to swallow. The characteristic warts on the skin of toads are the sites of swollen poison glands. In the marine toad there are two large glands on the back of the neck which produce a milky, poisonous secretion. So when the toad is attacked it is not its bite that is fatal but the white fluid exuded from its warts. This toad has few enemies probably because of the bad taste of this exudate but predators such as dogs and snakes are often killed by mouthing these toads. The toad blows itself up inside the throat of the predator, so suffocating it.

Murderous intruders

In 1863 sugar cane was planted in Queensland, Australia, near the town of Brisbane, and the first processing mill was set up. Because of the success of this venture more land was cleared for sugar, with losses among some of the native animals robbed of their habitat. Others benefited because cultivated land suited them. After 90 years the marine toad was imported from Hawaii, where it had been introduced in the hope it would clear the cane beetle. This was the hope also in Queensland, but now after 30 years, the cane beetle is still there, the toad has greatly multiplied and is spreading rapidly. Already it is widespread over a 1 000-mile coastal strip.

Toads do more good than harm, as a rule, but this is not the case with the marine toad. It is having a drastic effect on the native fauna. The smaller species have suffered in numbers from being eaten by the toad, while snakes and birds like the ibis were killed off when they tried to eat the toad, either by being poisoned or by having the toad impacted in the throat. An unexpected damage has been caused by the toad in dry summers when ponds and waterholes dry up. Such water as was left became so packed with breeding toads and with their spawn that it became undrinkable. Finally, as the cane beetle was still rampant, insecticides were used, especially benzene hexachloride, with the inevitable result that more of the native animals were killed off.

Already Australians see in the millions of marine toads now living in Australia a parallel story to that of the rabbit. Because in their original range the toads extend from the tropical to the temperate zone of South America there is reason to fear that in time they may spread throughout Australia. In addition, the marine toad is fast becoming a favourite laboratory animal. It is easy to keep and each year now more and more is being published on their physiology than on their habits in the wild. With such an adaptable animal, capable among other things of going without food for 6 months, it needs only a few to escape from laboratory animal houses for the toad to become established in unexpected places.

class	**Amphibia**
order	**Salientia**
family	**Bufonidae**
genus & species	***Bufo marinus***

Gathering of marine toads — a typical scene in the Queensland bush. These large toads are becoming a problem in Australia because of the rate at which they are spreading throughout the country and the effect they are having on the other fauna. The cane beetle, which it was hoped the toad would clear, still abounds!

Markhor

To many zoologists and sportsmen, the markhor is the most splendid and imposing of the wild goats. A male reaches 3 ft at the shoulder, occasionally 40 in., and weighs 180−220 lb; females are smaller, not quite so tall and only half the weight. The male has fine spiral horns, flattened and ribbon-like in some individuals, keeled both in front and behind, and divergent. The straight length of these horns varies from 20−40 in.; taken along the spiral, they are 30−65 in. long, and the tip-to-tip distance is from 8−44 in. Females' horns are not more than 10 in. long.

The markhor is coloured light foxy to sandy in summer, grey in winter, with a white-grey belly and blackish tail. There is a mane on the throat and breast which reaches a foot long in winter; and a mane on the neck and back which is half as long. The beard, present in both sexes, reaches 8 in. The name markhor seems to be derived from the Persian for 'snake-eater', but there is no evidence that they eat snakes.

The markhor lives in the precipitous mountains of the Hindu Kush and its outliers, the Kughitang and Kuljab in Turkmenia, the Suleiman range in Pakistan and the Astor, Gilgit, Pir Panjal and Chitral ranges in Kashmir.

Goats of the lower heights

Markhors live on the slopes of deep rocky gorges, where their agility and sure-footedness stand them in good stead. The markhor is not found at such high altitudes as the ibex. In Kashmir, for example, there are Siberian ibex as well as markhor; although the two live in the same ranges, markhor tend to keep to lower altitudes. Generally they are found between 4 500 and 8 000 ft. In summer some individuals, mainly adult males, go higher into the true alpine zone, at 10−12 000 ft.

Rutting by the calendar

For most of the year, markhor herds are very small − seldom more than five animals − and consist entirely of either males or females. In the breeding season, from November to December in the Kughitang range, herds of up to 27 are formed. The males fight at this season, rearing up and clashing their horns together like other goats. Gestation is 6 months, so the kids are born at the end of April or the beginning of May. The female leaves her group to give birth, then rejoins it when the kid can travel and keep up with the adults.

Several writers have testified to the regularity with which markhor in the Gilgit and Chitral come down each year from the higher regions of the Hindu Kush onto the lower spurs for the rutting season. On perhaps December 12 and 13 only females will be seen in the nullahs. Then, on December 14, whatever the weather, the 'big heads', as the males are called, appear as if by magic and the rut starts.

Toni Angermayer

A magnificent upstanding animal, the markhor has a heavy mane of long shaggy fur and splendid, greatly twisted horns which may be open as seen above or tightly closed.

Distinguished enemy

The land of the markhor is also snow leopard country, and this magnificent cat undoubtedly kills many markhor. At somewhat lower altitudes the true leopard, wolves and probably bears also make inroads into the population. Birds of prey, foxes and perhaps the manul, or Pallas's cat, take young ones. The manul, the size of a domestic cat, with soft pale fur, lives among the rocks, feeding mainly on pikas. Finally, the human element must not be discounted. In former times, many a sahib on the Northwest Frontier has longed to get that splendid pair of horns for the wall of the officers' mess.

Many twists and turns

Because they live high up in the inaccessible mountains, markhor are not among the best-known of the family Bovidae. Although it is so distinctive the species was not made known to science until 1839. Its common name is from the Persian *Mar* (snake) and *Khor* (killer) but whether it does kill snakes, and if so how, is not known.

Numerous subspecies or races have been named, based almost entirely on horns, the shape of which can vary enormously. It has been said that the race inhabiting the Astor range has very divergent horns, and each forms a very open spiral of only one or two complete turns. At the other extreme occurs the markhor of the Suleiman range which is said to have horns that form a tight V and each is a close corkscrew-like spiral with three or four complete turns. Certainly the more open type of horns predominate in Kashmir and Turkestan, and the tight V's in the south, but this is by no means a rule, and 'Astor' and 'Pir Panjal' types crop up way outside their supposed ranges. Indeed, there is so much variation in the horns that they are hardly a reliable feature on which to base nine races.

class	**Mammalia**
order	**Artiodactyla**
family	**Bovidae**
genus & species	*Capra falconeri*

Swiftest hunter: with spear-like bill and streamlined body a marlin executes a magnificent jump off La Paz, Baja California, being able to achieve speeds of up to 50 mph or even more.

Marlin

A fish combining grace with power and size, the marlin is among the most popular gamefishes. There are half a dozen species, including the spearfishes. The body is long and flattened from side to side, and the snout and upper jaw are drawn into a slender beak which is round in cross-section. The dorsal fin is low and continuous in young fish but with age the front part increases in height and the first few spines become greatly thickened. The anal fin is divided into two parts and the pelvic fins are at first longer than the pectorals but become relatively shorter with age. The tail fin is strongly forked, and there are keels at the base of the tailfin. The back is bluish, dark brown or black and the underside silvery, silver-grey or yellow. In some species the back and flanks are marked with narrow blue or silver bands.

The real number of species is uncertain but it includes the Atlantic white marlin, up to 9 ft long and 106 lb weight with a record of 161 lb, the blue, striped and black marlins of the Pacific, up to 14 ft long and 1 560 lb weight, the Cape marlin of South Africa, and the spearfishes. These last include the short-billed and long-billed spearfishes, of the Pacific and Atlantic respectively, and the Indian spearfish, or goohoo, of the Indian Ocean and Malay Archipelago, up to 6 ft long and 60 lb weight.

Ocean sprinters

Marlins are very powerful and are probably the fastest of all swimmers. They can reach speeds of 40–50 mph or even more. Such speeds are possible because of their shape; the body is streamlined, with the beak forming a highly efficient cutwater, and when the marlin is going at full speed all the fins, apart from the tail fin, are folded down into grooves in the body so there are no obstructions to an easy passage through water. Having the pelvic fins far forward, on a level with the pectoral fins, means that marlin and spearfishes can turn suddenly in a tight circle. Rapid movement through water places a great strain on the skeleton, especially when the fish has to brake suddenly. The backbone is made up of relatively few vertebrae and each of these has flattened interlocking processes that give strength and rigidity to the whole body.

Some indication of the speed and of the thrust of a marlin is seen in the way the beak has at times been driven through the timbers of ships. One whaler had a spear from a marlin or a spearfish that was driven through $13\frac{1}{2}$ in. of solid timber. In another case a 'spear' was driven through 22 in. of wood. There has long been speculation whether such incidents are due to accidental collision or deliberate attack. The former seems more likely.

Marlins lead solitary lives, well scattered about the ocean. In spring and summer they form pairs, so this is probably their mating and spawning time.

Thrust without parry

The main food of billfishes, as marlins and spearfishes have been called, is other fishes, especially mackerel and flying fishes, but squids and cuttlefishes are also eaten. The billfishes pursue the shoals for days on end, striking to left and right with the beak and then feeding at leisure on the dead and injured victims. It seems that the billfishes pursue, then stop to feed, overtake once more to take their toll, stop again to feed, and so on. A fish which moves at great speed needs a large amount of food to supply the necessary energy for this. Moreover, the muscular action generates heat and tests showed that the temperature of a five-striped marlin, about 9 ft long and weighing nearly 300 lb, was up to $6°C/11°F$ higher than the surrounding water. The marlins were tested with a thermopile harpoon—a harpoon carrying a device for registering temperature. They were played for a half hour or more before being landed, which would drive their temperature up, but it means that any billfish swimming fast has temporarily slightly warmed its blood.

Caught when exhausted?

It is something of a surprise to learn that the only enemies of these fishes are large sharks, especially the tiger shark and the maneater. Sharks are not especially fast swimmers. The blue shark has been estimated to reach $26\frac{1}{2}$ mph and the mako at 35 mph is probably as fast as any. Perhaps a comparison can be made between the fastest land animal, the cheetah (p. 550), which can keep up its great speed for only a few hundred yards, and the fastest fishes, the marlins and spearfishes. Possibly they also only use speed in bursts and must then recover, and this would be the time when they are vulnerable to the slower but relentless sharks.

Spearing their food

Clearly, to try to probe the secrets of an oceanic fish moving at speeds of 40–50 or more mph offers unusual difficulties. This has led to arguments in the past among deep-sea anglers as to whether marlins and other 'billfishes' ever spear their prey. In 1955 the *John R Manning*, longline ship of the US Fish and Wildlife Service, caught a marlin south of Hawaii. It weighed 1 500 lb and in its stomach was a freshly dead yellowfin tuna, 5 ft long and weighing 157 lb. The tuna had been swallowed headfirst and it had two holes through the body that corresponded with wounds from the marlin's spear. This more than justifies the name 'marlin', which is a shortened form of 'marline spike'.

class	Osteichthyes
order	Perciformes
family	Istiophoridae
genus & species	*Makaira albida* white marlin
	M. brevirostris short-nosed spearfish
	M. mitsukurii striped marlin
	others

Marmoset

Marmosets, the smallest of all monkeys, are South American. They differ from the more familiar monkeys in having claws, and in lacking wisdom teeth, the third molars. These, with its brain, are thought to be primitive features. For these and other reasons they are put, together with the tamarins, in a family separate from other South American monkeys.

Most marmosets are about 8 in. long, with a tail about 12 in., but the pygmy marmoset is under 6 in. long, with an 8 in. tail, and is the smallest monkey in the world. The lower incisors of marmosets are almost as long as the canines and are used in grooming. The common marmoset, from the Brazil coast, has silky fur, marbled with black and grey, a black head and long white tufts of hair around the ears. There are three or four closely related species from the Brazil coast, as far south as the Paraguay border, and the Amazon River basin. The pygmy marmoset, of the Upper Amazon, has no ear-tufts and the hair of the head is swept backwards over the ears. It is brown, marbled with tawny.

Males fight over territory

Marmosets are very active animals, bounding along branches, scurrying like a squirrel, with very jerky movements and helped by their claws. Family groups, made up of a mated pair and their offspring, live in the upper canopy of the trees, feeding on insects, fruit and leaves. For a long time there was some doubt whether the pairs were territorial because studies were largely made of captive animals and any fighting seen might have been due to incarceration. It is now known that the territorial instinct is strong and females are more aggressively territorial than the males. When two male marmosets meet, they threaten each other with rapid flattening and erection of the conspicuous white ear-tufts.

Father carries the burden

Breeding may take place at any time of the year. Unlike many other higher primates, marmosets have a courtship display. The male walks with his body arched, smacking his lips and pushing his tongue in and out. The two lick each other's fur, and groom each other using their long lower incisors as a comb. As in other monkeys the female menstruates at approximately monthly intervals. When she is in season the male is very active, marking objects by pressing the glands on his scrotum against them. Gestation is only 20 weeks. Twins are the rule rather than the exception among marmosets; in two-thirds of the births of common marmosets, and in 90% of those in pygmy marmosets, there are twins. Triplet births are commoner than single births. The male usually carries the young about; they cling to his back. The female may carry them but as a rule she has them only at feeding time. This caused a tragedy the first time marmosets were bred in captivity. The laboratory workers decided to take the male out 'to be on the safe side', so they removed the animal that was not carrying the young, that is, the female. The youngsters starved to death. Young marmosets are completely independent at 5 months.

Lurking enemies

There are several small cats in the Brazilian jungle that might kill marmosets as well as some birds of prey. A tame marmoset showed no fear of snakes when presented with either a live snake or a rubber model. By contrast, a tame marmoset fell to the ground in a state of seeming terror at the sight of a polished tortoise shell. This sheds no light on the marmoset's enemies except in comparison with certain African monkeys, which have been seen to fall out of trees in a dead faint at the sight of a leopard's eyes shining through the foliage.

With immense plumed topknot of pure white fur that looks like ostrich-feather headdresses worn by African chieftains, the cotton-topped tamarin **Sanguinus oedipus** has the most exaggerated headgear.

roebild

Primate Pygmies

Marmosets with the tamarins are the dwarf monkeys of the Amazonian region with characteristic shrill twittering voices.

▷ *The most brightly coloured of all living mammals, the golden lion marmoset sets a problem as no full explanation can be given why his coat is such an intense shimmering golden yellow. The most exotic of the marmoset species, the first living specimen seen in Europe was apparently owned by Madame Pompadour.*
▷▷ *White-headed sakis* **Pithecia pithecia** *grooming, an important part of social life.*
▽ *Pygmy marmoset: smallest of all living monkeys, its adult body length is only 4 in. For many years it was thought to be a juvenile form of some other marmoset species. When it was definitely established as a distinct type the dealers were unscrupulous, offering young common marmosets as the rare pygmy species.*
▽▷ *White-eared marmoset* **Callithrix aurita** *is one of the plumed marmosets. The small face has a triangular light blaze on the forehead and largish ears partly concealed by plume-like hairs that sprout from the cheeks.*
▽▷▷ *Silvery marmoset* **Callithrix argentata** *with zoo bred youngster — quite a catching colour with the glistening hair of the adult's black face and the baby's bright pink face and ears. As in other marmosets the male assists at birth, receiving and washing the newly born young. The father transfers the young to the mother at feeding time then accepts them from the mother again after feeding, often wearing them like a scarf round his neck.*

Zool Soc London

HE Uible: Photo Res

Zool Soc London

Mid-day 'hibernation'

Some tree-living animals, such as squirrels, have a group of long bristles on each wrist. These are tactile hairs (organs of touch) and marmosets are the only members of the higher primates to have them. This is one more primitive character to add to those already mentioned at the beginning. Perhaps the most primitive feature is the body temperature, which may vary by as much as 4 Centigrade degrees, from one part of the day to another, and is lowest about midday. This suggests that the marmosets have a period of torpor at the time when most other mammals rest because the sun is then at its hottest. A related South American monkey, the douroucouli, has a very even body temperature which does not vary by more than 1 Centigrade degree throughout the day. Moreover, the douroucouli maintains a steady body temperature when the surrounding air drops to as much as 8°C/46°F at night. A widely fluctuating body temperature is a feature of the lower vertebrates, such as reptiles, and when we find it in a mammal this almost certainly indicates that that animal is primitive.

class	**Mammalia**
order	**Primates**
family	**Callitrichidae**
genera & species	***Callithrix jacchus*** *common marmoset* ***Cebuella pygmaea*** *pygmy marmoset, others*

Jacana

Zool Soc London

Marmoset rat

*Although never called a bamboo rat, the marmoset rat's very existence depends on a single species of bamboo, **Gigantochloa scortrchinii**, found in southeast Asia.*

The marmoset rat is 6 in. head and body length with a tail 7½ in. long. Its plump body is covered with a soft silky fur, which is greyish brown turning to foxy or chestnut red and moulting back to greyish-brown on top with white underparts. The eyes are large and black, ears are large and nose blunt. The tail is scaly with only scanty hairs except in the end third which is hairy and something of a 'bottlebrush'. The feet are short and broad, pink and coated with white hair. Each toe, long and broad-ended, has a sharp claw at the tip except for the first toe of the hindfoot, which has a small nail. This toe is opposable to the rest. The rat can grasp bamboo stems by gripping with both first and fifth toes opposed to the three central toes.

The marmoset rat is known in the Malay Peninsula, Tenasserim, Thailand, Indo-China, and Hainan.

A bamboo home . . .

The marmoset rat can climb the slippery surfaces of the stout stems of bamboo by spreading its toes wide, and does so steadily and without difficulty, the claws taking no part in the action. It can also stop and stand or turn at any angle. On a stem leaning from the vertical it grips with the soles of the feet opposed, as a human climber might grip a pole, and moves in short spurts or bounds, moving alternately both forefeet and both hindfeet.

The naked skin on the undersides of the feet also helps in climbing. As with all rats there are pads on the soles, and there are narrow transverse ridges on the undersides of the toes. While the animal is active, the skin, especially that on the pads of the toes, is moistened by a sticky secretion that gives extra grip.

A marmoset rat enters the stems of a bamboo to nest by gnawing a hole about 1¾ in. diameter. Once inside it may use only the one cavity between the divisions across the stem or it may gnaw through them to occupy several cavities, and may at times make an exit hole, gnawed through from the inside. The nests themselves are made exclusively of bamboo leaves.

The structure of the stems of the bamboo *Gigantochloa scortrchinii* seems to be the reason for the rats' choice. In other species of bamboo growing in the same area the walls of the stems are thin and the fibres concentrated at the surface. This not only makes the surface difficult to penetrate but gives the walls a tendency to split longitudinally, and they break easily under stress when holes are cut in them.

. . . and bamboo meals

In captivity a marmoset rat was offered fruit to supplement its diet. It took a little papaya, banana and sweet potato. When it had learned, however, to drink from a water

Arboreal acrobat—the marmoset rat with naked, grasping feet can climb bamboo extremely well.

Jane Burton: Photo Res

bottle it rejected these, which suggests that the fruits only attracted it for the water they contained. It would also take paddy but there was no doubt that it preferred the growing tips of bamboo twigs, and especially fruiting twigs, and while it remained healthy on a diet of paddy and bamboo, if deprived of the latter it rapidly lost weight and condition. From another line of investigation it seems that its diet is even more exclusive. Examination of the stomach contents of rats taken in the wild showed large quantities of pollen and, while there could not be absolute proof of this, the pollen was probably bamboo and almost certainly that of the one species of bamboo. Many species of bamboo in southeast Asia flower rarely, some only at intervals of several years. *G. scortrchinii*, however, flowers and therefore produces pollen frequently.

Not so rare

Nothing is known of the marmoset rat's breeding habits or its enemies. This is not surprising because until recently it was an almost unknown animal, and believed to be very rare. At first it was known as Berdmore's rat, after Major Berdmore, who in 1859 collected a single specimen from Tenasserim, which was given the scientific name

Hapalomys longicaudatus, meaning the long-tailed soft mouse. In 1915 another specimen was taken at Pahang, in Malaya, and in 1927 a third, in Laos. In the same year a specimen collected in Hainan was described by the American zoologist AM Allen as *Hapalomys maronosa,* and given the colloquial name of marmoset mouse, which could be an allusion to its climbing habits and its general mouse-like appearance, although it is rat-sized. In 1956 another was trapped at Selangor, in Malaya, but so few specimens taken over such a wide area of southeast Asia, and over so long a period of time, suggested a rarity. Then Lord Medway published his account of the marmoset rat, as he prefers to call it, in the *Malayan Nature Journal,* for August 1963. In January of that year he found an area in Ulu Kelanton District where marmoset rats are common and well known. In fact, he found that the local Temiar regularly eat these rats and are skilful at hunting them.

If the bamboo dies?

It is, perhaps, this linkage with an exceptional bamboo which often flowers that sets the adaptations of the marmoset rat in a class apart. There are other animals that use bamboo for one purpose or another and have their particular specialisations, like the bat that has suckers on its wrists and feet, used in clinging to the shiny surfaces of stems. There is also a near relative of the marmoset rat with similar habits but not such close specialisations which offers a more useful comparison. This is the pencil-tailed tree mouse *Chiropodomys gliroides.* It also frequently gnaws holes in bamboo stems, to enter and nest in the cavities. Its holes are only 1 in. in diameter and are further distinguished by lacking a border of stripped skin, which is a characteristic feature of the holes made by the marmoset rat. The tree mouse, among other things, makes its nests mainly of leaves from broad-leaved plants, and is also less specialised in other ways. It is the more abundant in those localities where the two species of rodents occur together.

Intense specialisation leads to insecurity. This is a law of living organisms, for if anything occurs to upset the environment they are unable to adapt readily to the changed circumstances. The marmoset rat offers us an exceptionally clear example of the truth of this. If in any region the bamboo *Gigantochloa* were cleared, or for any reason died out, the marmoset rat would disappear with it. The tree mouse, on the other hand, would be able to adapt to the changed circumstances, or would stand a very good chance of doing so, because it is not wholly dependent on one species of bamboo.

class	**Mammalia**
order	**Rodentia**
family	**Muridae**
genus & species	***Hapalomys longicaudatus***

Hoary marmot usually lives below the timber line in Alaska either among rocks or in mountain meadows. Its sleeping nest of dry grass is deep in the burrows which may be several yards long and have many entrances so a disturbed family can soon retreat to safety below the surface.

Marmot

The number of species of marmots is uncertain; estimates range from 8 to 16. Among the best known of them are the alpine marmot and the bobak, both of Europe, and the hoary marmot of North America; there are other alpine species in both Old and New Worlds. Equally well known is the woodchuck of North America, a marmot differing markedly in its habits, which will be dealt with later.

The alpine marmot is up to 2 ft long with a 6in. bushy tail. It weighs up to 18 lb. Stout-bodied with short legs, its head is wide and short with small rounded ears and large eyes. The fur is coarse and stiff. It lives in and around the treeline, from about 4 to 9 thousand feet, in the Alps and Carpathians and in corresponding alpine districts of central and north-eastern Europe. The bobak, Himalayan marmot or steppe marmot is similar to it in life history but lives more in virgin grasslands. Because agriculture has taken much of its habitat, it has become extinct over large areas of its former range,

which included central and southern Asia and extreme eastern Europe. The bobak is golden brown with a black tail tip. The alpine marmot is pale brown with its back and crown peppered with black, white markings on the face and the outer half of the tail black. The hoary marmot, of about the same size, is silvery grey peppered with black on the back and rump; the face is black with white cheeks. The forehead, lower legs and feet are black. It lives in mountainous parts of western North America, from Alaska south to New Mexico. Its habits and life history are similar to those of the alpine marmot.

Mountain shelters

Marmots leave their burrows on sunny, boulder-strewn slopes, as the first rays of the rising sun fall on them. When an entrance is shaded there may be some delay, and individual marmots differ, some seeming more reluctant than others to leave their beds. The sleeping nests of dry grass are deep in their burrows, which may be several yards long and which have several entrances. In autumn they change the grass in the sleeping chambers and in October the marmots hibernate, the whole family sleeping together with the burrow entrances blocked.

Two meals a day

The first hour after dawn is spent in sunning and grooming near the burrow entrances, after which the marmots disperse to the feeding grounds, walking with tails held in an arc with the tip towards the ground. They feed on grass, sedges and herbaceous plants as well as roots. They have no cheek pouches for carrying food like so many of the related ground squirrels. Feeding lasts for about 2 hours, after which the animals resume their grooming and sunning, with some digging. There is a further feeding period in the 2–3 hours before sundown before they retire for the night.

Slow growers

Breeding begins when hibernation has finished and the bedding in the nesting chambers has been renewed. The gestation period is about 42 days and there are usually 2–4 in a litter, born in the early days of June. The babies first come out of the burrows in mid-July. They stay with the parents until the following spring. It is not certain whether a female breeds every year or every two years. The young are slow to develop; they do not reach full size until 2 years old, and probably do not breed until 3 years of age. Alpine marmots have lived for 20 years in captivity.

Joe Van Wormer: Photo Res

Sounding the alert

Marmots are too large for the smaller flesheaters such as stoats. Their main enemies are foxes and eagles. Even these probably take relatively few. Moreover, marmots are alert to danger and very wary, and they are famous for their alarm whistle. They have, in fact, been credited with posting sentinels, whose whistles send all other marmots within earshot scurrying to their burrows. One suggestion is that it is usually an elderly female who acts as sentry for the colony, posting herself on a high boulder while the rest are feeding, and giving a shrill whistle whenever danger threatens. It seems, however, that marmots have a low barking call, a low whine and a whistle, by which they communicate with each other, the whistle being used whenever they are ill at ease. It is used, for example, by a marmot that ranks low in the 'peck order'. Babies have a shrill squealing very like the adult whistle to attract their mothers' attention. Whenever a large animal appears within sight of a colony the marmot nearest to it whistles. The nearer the intruder comes, the more shrill the whistle. Marmots feeding near the centre of a colony are less prone to whistle. So it follows inevitably that individuals on the fringes of the colony are the most likely to sound the alarm, so giving the impression to any person drawing near to the territory of a colony that there is a sentinel giving the alarm.

Marmots make hay

Another ancient belief, one which has persisted for 2 000 years, is that marmots will carry grass to their burrows much as rats are reported to transport eggs (see p. 633). That is, one lies on its back while the other loads hay onto it then pulls it along by its tail. Zoologists are highly sceptical of this story, yet even within recent years letters have appeared in Swiss newspapers from people who claim to have witnessed it and giving very precise details of how it is done. The whole story seems highly improbable but so many remarkable things are being brought to light by the use of ciné cameras and other sophisticated techniques that it would be unwise to say such a thing as marmots playing hay-wains does not happen. We can only reserve judgement.

class	**Mammalia**
order	**Rodentia**
family	**Sciuridae**
genus & species	***Marmota bobak*** bobak **M. marmota** alpine marmot others

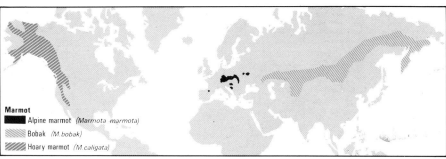

Marmot
- ▨ Alpine marmot *(Marmota marmota)*
- ▨ Bobak *(M. bobak)*
- ▨ Hoary marmot *(M. caligata)*

◁ △ *Displaying its name to perfection, the yellow-bellied marmot is smaller than the hoary marmot. Although in a similar range it lives lower down the mountains and always locates its burrows in the rocks.*
▷ *Alpine marmots in typical feeding stance.*

Constance P Warner

Constance P Warner

Marsupial frog

There are several species of South American tree frogs the females of which, instead of staying with the spawn to guard it, like some frogs, carry the spawn about with them in pouches. The first scientists to discover this, in 1843, merely saw that the female had a pouch and gave the frogs the scientific name **Gastrotheca.** *Translated literally this means 'stomach pouch', but in fact the pouch is on the female's back.*

Marsupial frogs are fairly ordinary tree frogs except that some are very small, $\frac{3}{4}-1\frac{1}{4}$ in. long. The largest are only 4 in. long. They have sucker pads on the tips of the toes, usual in tree frogs, and some have stouter legs than most tree frogs. Their colours are mainly green with brown spots, blotches or stripes.

The tadpole-pocket

Marsupial frogs have an unusual way of taking care of their young, which varies from one species to another. In the smallest of them each female lays only 4–7 eggs, rich in yolk. The larger species may lay 50 or more. In these the eggs remain in the pouch until they hatch and the tadpoles stay there also, leaving the pouch only when they have changed into froglets.

When these frogs are pairing the male clasps the female as usual but he is slightly farther forward on her back. Just as she is about to lay, the female raises herself on her hindlegs so her back tilts steeply downwards towards her head. Her cloaca, the opening through which the eggs are laid, is directed upwards. As a result the eggs roll down her back and into the pouch, the male fertilising them as they go. Once the eggs are all safely inside, the mouth of the pouch closes and the surface of the female's back looks lumpy. How the froglets get out was discovered about 15 years ago; the female lifts a hind toe over her back and

△ *Tadpoles-in-waiting: female marsupial frog* **Gastrotheca mertensi** *with young in back pouch to which eggs were transferred on laying. As tadpoles they escape from the pouch singly into the surrounding water (above right).*
▽ **Gastrotheca marsupiata** *— the generic name means 'stomach pouch'.*

Peter Livesley

pulls apart the edges of the slit-like opening to the pouch.

As adults the frogs live in trees, eating insects. Because of their unusual baby care, they do not have to go to water to spawn.

Launching the tadpoles

Several species of marsupial frogs let their young go as tadpoles; the female goes to water and lowers herself into it. Then she brings her hindlegs over her back and puts the first toe of each foot into the opening of the pouch and pulls, so the mouth of the pouch opens wide to let the tadpoles out. They come out one at a time, at short in-

tervals. These species also have a different way of putting eggs in the pouch. They lay up to 200 eggs and as the spawn is being given out by the female the male fertilises it as usual. At the same time he uses his hind feet to push the spawn into the pouch.

Inside the pouch the tadpoles are enclosed in the egg membrane. Their external gills are much enlarged to form a sort of placenta for breathing. The details of how this works during the 100–110 days inside the pouch are not yet known, but soon after the tadpoles leave the pouch the gills are quickly absorbed. The hindlegs appear 26 days after the tadpoles enter water, the forelegs grow out 19 days later and at the end of 56 days from the time of release, they have changed into froglets.

The incredible frog

The discovery of a frog that brooded her eggs in a pouch was remarkable but over 100 years passed before it was known how the eggs got into the pouch. Then in 1957 Professor EC Amoroso of the Royal Veterinary College in London, having some marsupial frogs sent over from South America, filmed the whole process, giving zoologists everywhere a chance to see a quirk of nature which might have been scorned by the scientific world if presented in another way. But there is more to come; the discovery of how the tadpoles got out of the pouch. Even this is not the end because apparently the female marsupial frog can feel when an egg is about to burst and release the tadpole. She has been seen to flex one hindleg over her back to let a tadpole out of the left side of the pouch, then flex her right leg over her back to release a tadpole from the other side.

class	**Amphibia**
order	**Salientia**
family	**Hylidae**
genus & species	***Gastrotheca marsupiata*** *others*

Marsupial mole

The marsupial mole looks very like a true mole except the female has the usual marsupial pouch for carrying her babies. It used to be thought there were two species but the latest view is that there is only one. The marsupial mole has a cylindrical body, short neck and legs and is covered with long, silky fur. Usually only 3½ in. long, it may grow up to 7 in. There is a horny shield covering the snout. The only sign of an ear is a tiny hole each side of the head covered with fur. The vestigial eyes are covered by fur and have no lens or pupil and only a weak optic nerve. Each forefoot has a cloven scoop formed by the curved and much enlarged claws on the third and fourth toes, the other three toes on the forefeet being small. The middle three toes of the hindfeet also have slightly enlarged claws. The hard, leathery tail, ½–1 in. long, is marked with rings and knobbed at the tip.

The size seems to vary from one part of the animal's range to another. It is smallest in northwestern Australia. The range of the marsupial mole is from the deserts of south-central to northwestern Australia.

The marsupial mole is very like the true moles of the Northern Hemisphere but for the pouch.

Australian News & Information

Why our ignorance?

Nothing is known about their breeding habits or their enemies. Ellis Troughton tells us that the first specimen of a marsupial mole to be brought to the attention of scientists was caught by William Coulthard, manager of Idracowra cattle station. This included several hundred square miles of country west of the telegraph line between Charlotte Waters and Alice Springs. Coulthard saw a most unusual track in the sand near his camp, followed it and found the mole under a tussock of spinifex, or porcupine grass. Although he had had long experience of that area, this was the first time he had seen either the track or the animal. AG Bolam, stationmaster at Ooldea on the trans-Australian railway, was a keen naturalist who caught a few of the marsupial moles that have been kept in captivity. Yet he was unable to add much to the little that was known. The first specimen to be captured in northwestern Australia was in 1910 and then a young specimen was captured in 1946. As a rule marsupial moles have been brought in by Aborigines.

Evolutionists' dream

We are told that the first marsupial mole caused as much excitement among scientists as did the first platypus. By 1888 the furore caused by the publication of Darwin's *Origin of Species* in 1859 had hardly abated. Scientists were still searching for clues to support the general idea of evolution when suddenly

Australian News & Information

Iain Macmillan at Nat Hist Museum

Keeps out of sight

The marsupial mole is a relatively recent discovery; the first one was not found until 1888. This is because it does not make permanent burrows, so it is far more difficult to find than the true moles with their molehills and permanent underground systems. After rain a marsupial mole comes onto the surface and leaves a characteristic triple track, made by the two forepaws and the stumpy tail. This sometimes leads a naturalist to a marsupial mole before it has dived into the sand where it burrows about 3 in. below the surface. These moles are, however, not rare; when the railway was built from Perth to Port Augusta, many were unearthed—but not studied.

△△ *Lower view sketch of Australia's underground earth-mover. Note the outsize claws.*
△◁ *Side view sketch, showing the tiny eye and the reinforced snout (inset).*
△▷ *A museum specimen, showing colour.*

Making the most of life

Marsupial moles have the same restless and feverish way of life as true moles. After eating, they fall asleep suddenly, breathing rapidly as if trying to concentrate as much rest into the time as possible. They wake up, equally suddenly, and start their feverish search for food again, as if trying to crowd as much eating into the time as possible. In captivity marsupial moles eat earthworms and insects ravenously.

they were presented with this first-class example of convergent or parallel evolution. Except for the colour of their fur the two moles looked almost exactly alike. Yet they could not be related as one had a pouch and they lived poles apart. It must therefore have been because they lived in the same way and had the same habits, that they had come to look like twins.

class	**Mammalia**
order	**Marsupialia**
family	**Notoryctidae**
genus & species	***Notoryctes typhlops***

Marsupial mouse

Marsupial mice belong to the same family as the Australian marsupial carnivores such as the native cats, Tasmanian devil and the thylacine or Tasmanian wolf. These are much larger than the marsupial mice and all have a reputation as bloodthirsty killers, and in their smaller way the marsupial mice share this quality.

There are a score of mouse-sized marsupials that are variously known as marsupial mice or marsupial jerboas, depending on whether their hindlegs are long or not. They live in Australia except for two species, one in Tasmania and the other in New Guinea. All are delicately built with sharply pointed snouts, long whiskers, large ears and in most species, hindlegs not markedly longer than the forelegs. Their fur is reddish-brown to grey with white underparts. Although they look so much like mice they can readily be told apart from true mice; instead of two incisors in the upper and lower jaws, as all rodents have, marsupial mice have 8 incisors in the upper jaw and 6 in the lower jaw.

Some species, such as the broad-footed marsupial mice, range throughout Australia in suitable habitats, whereas others are restricted more to one part of the continent or another.

Living in scrub and desert

There are several different kinds of marsupial mice, each with its own peculiar shape or habit, reflected in the different names. Although called 'mice' they can grow up to a foot long, of which slightly less than half is tail. The 10 species of broad-footed marsupial mice, including the dibbler, live in rocky country with trees and undergrowth, especially near streams. They leave their shelters in rock crevices, hollow logs and caves at night and they climb well, into trees and vines. A widespread species related to the broad-footed marsupial mice is the flat-skulled Ingram's planigale, just under 5 in. long of which nearly half is tail. The marked peculiarity of this species is the flattened skull. In an animal $2\frac{3}{4}$ in. long in head and body the skull is $\frac{1}{6}-\frac{1}{4}$ in. deep. This seems a clear adaptation to creeping through crevices in rock in the rocky and sandy areas it inhabits. The crest-tailed marsupial mouse, $8\frac{1}{2}$ in. head and body and 5 in. tail, lives in deserts. It has very short legs and a tail that is thickened and well furred,

especially in the first half. This contrasts with the last group, the jerboa marsupials, which also live in the deserts of central Australia but have very long hindlegs and a long slender tail with a tuft of long hair at the tip. They spend the day in deep burrows or in hollow logs, coming out in the evening and remaining out until dawn, hopping about in leaps of up to 6 ft.

The flat-skulled Ingram's planigale is the smallest of the living marsupials in Australia. The largest of them is $2\frac{3}{4}$ in. head and body and 2 in. tail. The smallest of this same species, from the Kimberley District of Western Australia, is only $1\frac{3}{4}$ in. head and body with a 2 in. tail.

Mouse-eating marsupial mice

Marsupial mice probably eat any animal food small enough for them to tackle, and we are told by one Australian naturalist after another how they will eat the introduced house mouse. One marsupial mouse in captivity killed a house mouse in a flash. At sight of it the marsupial mouse suddenly went rigid, its tail quivering, then it leapt at the mouse, bit it in the back of the head, killing it instantly. Before going further it cleaned its whiskers and groomed itself—

◁ *Letting off steam. A vicious-looking broad-footed marsupial mouse emits a loud hissing noise during its threat display. Although very active they are shy creatures, rarely found near human habitation.*
▽ *A Macleay's marsupial mouse carefully*

Harry & Claudy Frauca

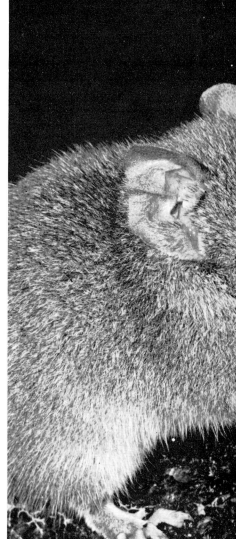

washing its hands before a meal!—then starting at the tip of the snout it methodically ate its way to the tip of the tail, turning the skin neatly inside-out as it went.

Another, which itself weighed only ¾ oz, ate a whole ounce of food in one night: 5 large insect grubs and 3 small lizards, the lizards being eaten bones, skin, tail and all.

All the marsupial mice seem to be carnivorous and much of their food is insects, although at least one kept in captivity is reported to have eaten cake. The full list of their diet is not unlike that of shrews and one of the discoveries in recent years, which seems to have escaped the notice of early naturalists, is that the common shrew must eat seeds or grain to keep healthy. The chances are, therefore, that marsupial mice also eat some plant food. More certainly we know they eat insects, including beetles, cockroaches and termites, as well as centipedes, spiders and small lizards.

Pouches and no pouches

Although we speak of these mammals as marsupial mice, not all have pouches. In some species it consists of nothing more than a fold of skin surrounding the teats, and in others it is a complete pouch, as in a kan-

prises apart the wings of an insect it has caught. Its needle-cusped teeth are especially adapted for an insectivorous diet although they will tackle almost any animal small enough for them.
▷ *Ears pricked, a bewhiskered narrow-footed marsupial mouse begs the question.*

garoo, except that in some marsupial mice, noticeably the jerboa marsupials, the pouch opens backwards. Even when the pouch is incomplete the babies remain attached to the teats for about a month—the same length of time as they stay in the pouch in other species. Some of the marsupial mice may carry up to 10 babies at a time. One female disturbed by a plough had 10 youngsters that had just left the pouch. They were clinging to her flanks as she tried to make her way over the uneven earth. The ploughman watching this removed the babies. The mother remained near and in response to their squeaks threaded her way among them until they had all taken hold of her fur once more and she could move to cover with them.

Marsupial mice, even those that are widespread, seem unable to survive near human settlements where domestic cats are kept or have gone wild. Elsewhere they are preyed upon by snakes, owls and their larger relatives the native cats (dasyures) and the Tasmanian devil.

Like flies on a ceiling

The broad-footed marsupial mice have grooves running across the pads on the soles of their feet which give them the grip necessary for climbing into trees and vines or over rocks. The toes also have long claws

which help. These mice make nests of gum leaves tightly packed or interwoven, in hollows or crevices or in abandoned birds' nests. The mice have been known to build in the old nests of lyrebirds, sometimes high up on cliff faces. They have also been seen to have nests in crevices in the ceilings of sandstone caves. To reach these, and also to carry materials to the crevices to build the nests, they run swiftly upside-down across the roofs of the caves, hanging on with feet that have sucker-like pads on the soles and long claws.

class	**Mammalia**
order	**Marsupialia**
family	**Dasyuridae**
genera & species	*Antechinus* *broad-footed marsupial mice* **A. stuartii** *Macleay's marsupial mouse* **Planigale ingrami** *Ingram's planigale* **Dasycercus cristicauda** *crest-tailed marsupial mouse* **Sminthopsis crassicaudata** *fat-tailed dunnart* **Antechinomys spenceri** *wuhl-wuhl, others*

Ederic Slater

Ederic Slater

Marten

Martens are members of the same family of mammals as the badger, otter, ferret and honey badger. They look like large ferrets, with long, lithe bodies, short legs and pointed muzzles. There are several species in both the New and Old World. The American marten of North America is 2 ft long, including a 10 in. tail with light red to black fur. The fisher or pekan is larger, 30—40 in. long, and dark brown or black. There are six species in Europe and Asia. The pine marten is 25—30 in. long and

Arboreal artist

Martens are excellent climbers, ascending and descending tree trunks headfirst and leaping through the network of thin branches with the ease of a squirrel. If they fall, they land feet first like a cat. There is a report of a sable falling out of an aeroplane while being transported during a recolonisation programme in the Soviet Union, and surviving the fall to be recaptured some months later.

In some places martens live on open rocky ground but even there they are not often seen. This is because they are solitary and mainly nocturnal and in many parts are becoming increasingly rare.

James Simon: Photo Res

weighs 2—3⅓ lb, the females are slightly smaller than the males. It has a bushy tail and broad, triangular head. The coat is a rich brown, darker on the middle of the back and the legs. The underparts are greyish and on the throat there is a cream patch which may have an orange tinge in winter. It is the upper long hairs of the fur that give the coat colour. Beneath these there is a short undercoat of reddish-grey hairs tipped with yellow. The pine marten still lives in the wilder parts of the British Isles, and extends across Europe and onto the plains of eastern Siberia. Farther east lives the closely related sable. The beech or stone marten is found from the Baltic Sea to the Mediterranean and eastwards to the Himalayas and Mongolia. The yellow-throated marten lives to the east of the beech marten.

Like all members of its family, which includes the notorious skunk, the pine marten has special scent glands at the base of the tail. The secretions from these glands are used to mark the home range as a dog marks its range with urine. Only the skunk and polecat use their secretions to deter their enemies. The odour of the pine marten is not objectionable and it used to be known as the sweet marten to distinguish it from the foulmart or polecat.

Enemies of squirrels

Pine martens chase squirrels through the trees, and it is probably the near extinction of the polecat and pine marten in England that allowed the introduced grey squirrel to colonise the country so quickly. They also hunt rabbits, rats, mice, field voles, small birds and occasionally game birds and poultry, and will eat caterpillars, beetles and bees, as well as blackberries, bilberries and

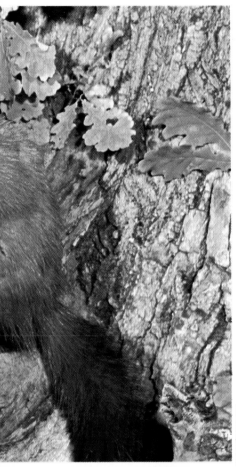

◁ *Still life. A 6-week-old pine marten crouches quietly in his grassy den. His safety lies in remaining unnoticed.*

◁ *Well wrapped for the winter. This yellow-throated marten has two fur coats: a shorter undercoat and a top coat of longer hairs which give the coat its colour.*

◁▽ *Cornered. A seemingly defenceless mother pine marten peers up as she lies curled round her young in their den in a hollow tree. She will become extremely savage if provoked.*

▽ *Snow trail. An American marten winds its way over the snow with its pointed muzzle eagerly searching for food. Although martens are active throughout winter, they often rest when the weather is really bad.*

3–4 young in each litter, but there may be any number from 2–7. They are born in a grass nest among rocks, in a hollow tree, or in the nest of a crow or squirrel. The young first come out of the nest when they are 2 months old and romp around nearby, producing a conspicuous 'playground' of flattened grass and herbs. They are weaned at 6 or 7 weeks and leave their mother soon after. Female pine martens are mature when 1 year old, produce their first litter when 2 years old and live to about 17 years.

The breeding habits of the American marten and fisher are similar, except that the American martens are weaned and become independent earlier. The female fisher

cherries. Slugs are eaten after the slime has been removed by rolling them under their paws and HG Hurrell has seen a tame pine marten striking down bumblebees as they visited flowers. Bees' nests are dug up for the grubs and honey.

The American marten has a similar diet to the pine marten but the fisher hunts more on the ground. Its name is a misnomer as it probably feeds only on spent salmon or steals fish from traps and nets. It catches beavers, American martens and smaller animals and kills porcupines by flipping them onto their backs. It has been known to kill deer that have foundered in deep snow.

Long gestation

Mating takes place in June or July but the young pine martens are not born until the following April as there is delayed implantation of the embryos (see armadillo p. 176 and badger p. 274). There are usually

leaves her litter, mates, then returns to them and rears them while the development of her new litter is delayed. Presumably other martens behave in the same fashion as the male does not stay with the female while she rears the young.

Easily caught

In the Middle Ages martens were abundant in Europe but they were trapped in such large numbers for their fur that they became rare. They were also trapped in the cause of game preservation in Britain. The last one to be killed near London was shot in Epping Forest in 1883, but since the Second World War two have been killed in southern England. In Siberia the sable was all but exterminated for the sake of its rich fur. It was discovered just in time that sable could be reared on ranches like mink, and since then they have been bred not only for their fur

but also for reintroduction into the places where they were exterminated.

In North America there was a similar story. Under the name of 'American sable' or 'marten' thousands of pelts were sold annually and the martens were exterminated in many places. Unfortunately, the American species do not breed readily in captivity. Protection on both sides of the Atlantic, however, has allowed numbers of martens to recover in some places.

Martens are very easy to catch which is probably the reason for the serious reduction in their numbers. They are attracted by bright objects and have other weaknesses, as one scientist discovered. While studying them he was able to catch the same animals repeatedly by baiting his traps with kippers!

class	**Mammalia**
order	**Carnivora**
family	**Mustelidae**
genus & species	***Martes americana*** *American marten*
	M. flavigula *yellow-throated marten*
	M. foina *beech marten*
	M. martes *pine marten*
	M. pennanti *fisher*

◁ *Barking up the wrong tree? A pine marten's attention is caught by something moving in another tree in the forest.*

▽ *Animal snow plough. A pine marten digging his way through the deep snow is abruptly interrupted at his work.*

Bavaria

△ *Cradle snatcher? A beech marten props itself up to enjoy the contents of a bird's egg, one of its favourite foods. These solitary creatures usually travel in a series of small overlapping areas in search of food, but they may travel up to 10 miles in a night.*

The two-tone effect of the coat is due to the lighter underfur showing through. This underfur may even be white in some beech martens. The beautiful coat of this animal has been its ruin, however, as fur trappers have seriously reduced the numbers of martens.

Marten
- American marten *(Martes americana)*
- Yellow-throated marten *(M. flavigula)*
- Beech marten *(M. foina)*
- Pine marten *(M. martes)*
- Fisher *(M. pennanti)*

Martial eagle

The martial eagle is the largest of the African eagles. Like many other eagles it bears a crest. It has long wings and a relatively short tail and in flight can be confused with only one other eagle, the serpent eagle. The upperparts are dark grey with light grey bars on wings and tail. The underparts, including the feather 'leggings', are white, barred and spotted with black. The bill is black and the legs and toes, which are armed with long curved talons, are blue-grey. The total wingspan may be as much as 8 ft. The females have larger spans than the males and are more powerfully built. They are easily distinguished, being more spotted on the underparts than the males.

The martial eagle lives in Africa from the southern borders of the Sahara to the Cape, but not in the thickly forested regions such as Zaire.

Shy eagle

A pair of martial eagles inhabits a range of as much as 50 sq miles, soaring over the countryside for hours at a time, often at great heights where they are almost invisible to the naked eye. Martial eagles are shy birds as compared with other eagles, and shun human settlements, which is to their advantage as they are often persecuted for taking farm stock. Because of persecution and their dislike of inhabited areas, martial eagles are much rarer than they once were. They are found in savannah, semi-desert and other

◁ *Do not disturb: a martial eagle discourages intruders on its reptile repast—monitor lizard.*
▽ *A golden glower from Africa's largest eagle.*
▽▷ *Grounded: sub-adult martial eagle showing its white chest and abdomen.*

open country, and breed only in forested regions when there is open country nearby.

Dropping in for a meal

Martial eagles spot their prey from a great height, swooping down on it in a well-controlled glide. The speed of the descent is regulated by the angle at which the wings are held over the back. When they are held almost horizontal the glide is shallow and the descent slow, but if the wings are raised in a 'V' they get less lift and the eagle drops at a steep angle.

They usually prey on small mammals and birds that live in the open, but the species vary from place to place. Their favourite foods seem to be game birds such as francolin, bustard and guinea fowl and mammals, like hyraxes. They will even eat impala calves. Jackals, snakes and lizards are sometimes taken but martial eagles rarely eat carrion. Domestic poultry, lambs and young goats are often eaten, but Leslie Brown, the authority on African eagles, has suggested that on the whole martial eagles are beneficial to man and that their destruction of livestock has been exaggerated.

Choice of nests

Martial eagles build large nests of sticks in tall trees, often on hillsides so there is a clear run-in to the nest. The female builds the nest, which may be 4 ft across and 4 ft deep, while the male collects sticks, or even small branches. The nests are used year after year and usually the female has to do no more than repair the nest and add a lining of fresh green leaves. Some pairs of martial eagles have two nests, each being used in alternate years.

Nest repair may take several weeks and when complete a single white or pale greenish-blue egg with brown markings is laid. The laying date varies between November in Sudan to July in South Africa. The female alone incubates and broods the chick

when it hatches out after about 45 days. For about 2 months the male brings food to the female who then gives it to the chick. Later, the female also hunts for food for the chick. The chick makes its first flight when it is about 100 days old. For some days it returns to the nest to roost and thereafter it stays fairly near the nest. Young martial eagles have been seen near their parents' nest when 3 years old. Unlike the crowned eagle that breeds in alternate years, the martial eagle may breed several years in succession, then fail to do so for several years.

Separate interests

In the course of his remarkable studies on African eagles, Leslie Brown found a hill on which five, and in one year six, species of eagles nested. The hill was appropriately named 'Eagle Hill'. There seemed to be no reason for this gathering except a natural gregariousness; there was no special abundance of food and there were plenty of other suitable nesting places nearby.

While on 'Eagle Hill' the different species did not interfere or compete with each other. The martial eagles fed on game birds caught in open country whereas African hawk-eagles hunted those in bush country. Brown snake-eagles caught snakes. Verreaux's eagles ate hyraxes that they hunted among rocks; the crowned eagles preyed on duikers and monkeys in the forests and Ayres' hawk-eagles took small birds from the trees. So, although crowded, the eagles did not have to compete for food.

class	**Aves**
order	**Falconiformes**
family	**Accipitridae**
genus & species	***Polemaetus bellicosus***

Peter Johnson

KB Newman

Martin

'Swifts, swallows or martins, they're all the same to us. We call them all swallows here.' These were the words of a countryman some 40 years ago. Nowadays people are better informed about birds. Some people still get these three birds confused as they are similar in appearance and habits, but swifts belong to a different family from the martins and swallows.

The house martin breeds in Europe, most of Asia as far east as Japan, and in North Africa. Its tail is less forked than a swallow's, the upperparts are blue-black, the underparts are white and it has a white rump. The latter provides the easiest way of distinguishing it from a swallow. Young house martins are browner than the adults, but are unlikely to be confused with the sand martin, which is smaller and more slender than a house martin and is brown above with a brown band across the otherwise white underparts. The sand martin ranges around the world. It is found in most of America north of the Mexican border and in most of Europe and Asia except the Orient. In Africa it is found from Egypt to Abyssinia.

Among other members of the swallow family which are known as martins are the American purple martin, the male of which is purple while the female is brown, the Mascarene martin of Madagascar, the crag martin of Africa, southern Europe and Asia and the tree martin of Australia. The river martin of Zaire, black with red eyes and beak, is sometimes placed in a separate family as it is rather different from other martins and swallows.

In 1968 the discovery of a new species of river martin was reported from a lake in central Thailand. This discovery is quite remarkable as the Zaire river martin, 5 000 miles away, was until then, the only known member of the genus **Pseudochelidon**. The shafts of the central tail feathers of the Thai species, **P. sirintarae**, which are only slightly prolonged in the African species, are about 3 in. long with racquet endings.

Down to earth. House martins flitting about in the mud have left the skies and are busy collecting material with which to build their nests.

Following the sun

The flight of martins is fast and twisting; a group feeding presents a high speed aerobatic display, as they seek midges, flies, beetles and butterflies. They rarely land on the ground, preferring to settle on telegraph wires or walls of buildings. As their diet is flying insects they migrate southwards from temperate regions in the autumn and return in the spring. Ringing migrant birds has shown where they migrate to and the manner in which they migrate, and many experiments suggest that they navigate by the sun. There are, however, many mysteries of migration still to be solved.

As the weather becomes colder, or food scarcer, flocks band together as they move south and finally leave in streams, travelling at 20–30 mph. Sand martins spend the winter in West Africa, from Gambia to Nigeria, and in eastern Africa from Kenya to South Africa. House martins migrate to Africa and India and other martins go similar ways to warmer climates. The brown-chested martin, for instance, nests as far south as Tierra del Fuego and migrates to the Amazon basin, Venezuela and the Guianas. In the following spring the flow of birds is reversed and they gradually move northward, or southward as the case may be, as the weather improves, arriving at their nesting sites in temperate regions some time in the spring.

Happy families

Martins nest in colonies and throughout the breeding season show a well-developed communal life. The nests may be mud or be in burrows in banks or sandy cliffs. The purple martin of America can be attracted to nest boxes and the brown-backed tree martin often uses the abandoned nests of ovenbirds. The house martin originally nested on cliffs, and still does in Japan, but now they build their nests of mud and bents under the eaves of houses. Gathering mud is a communal event, the house martins from a colony fly together to the muddy bank of a stream or pond and gather mouthfuls of mud, which are then stuck together and strengthened with bents to form the typical round nest. The final result looks as if it is made of round pebbles glued together. As well as gathering material communally, several martins may work on one nest. Sometimes the house martins need do no more than repair last year's nests, but at other times they may have to rebuild those that have fallen down or construct

Portrait of a sand martin. This brown and white swallow of the water, usually freshwater, catches most of its food as it skims low over the surface.

Heinz Schrempp

1575

Ronald Thompson

▷ *Approach route: a house martin carefully manoeuvres itself ready to land on its dome-shaped nest of mud and saliva.*

▽ *A mother sand martin cannot even get inside the front door before being greeted by her squawking, hungry offspring.*

▽ ▽ *The departure wires. Flocks of martins line up in orderly rows ready to migrate to a warmer climate. Although they often rest for short periods on wires, they spend most of their time on the wing.*

Eric Hosking

Fritz Siedel

completely new nests to accommodate a rise in population. Sand martins nest in burrows 2–3 ft long dug with their feet in banks, on cliffs, sandpits, railway cuttings or in rare cases inside drainpipes. The same burrows are used year after year.

Martins lay 4 or 5 white eggs which are incubated for a fortnight or so by both parents. The chicks are fed by both parents on regurgitated insects. The chicks stay in the nest for about 3 weeks before they take their first short flight.

Martins raise 2 or 3 broods each year and earlier families help to rear later broods. The first family of martins feed their younger brothers and sisters, and they all squeeze into the nest at night to roost. The record is 13 martins roosting in one nest. There were two adults, three nestlings and four young birds from each of the two previous broods.

Sparrows take over

The reactions of martins to hawks and falcons that come near the colony is to band in a group, mob the predator then retreat into their nests. There is, however, a more insidious enemy. House sparrows have the habit of taking over martins' nests, much to the annoyance of some people who prefer house martins to sparrows under their eaves. Usurped nests can be recognised by the untidy wisps of hay that hang out of the entrance. Often the sparrows come to grief because they cannot maintain the fabric of the nest like the rightful owners and the nest may suddenly collapse. There are also reports that the house martins sometimes plaster up the entrance, imprisoning the sparrows inside.

Maiden flight

The first flight of a young martin is often little more than a short flutter and glide to a convenient perch or down to the ground. There the fledgling may stay, apparently content, while several adults zoom over or land by it. They appear to be encouraging

Fritz Siedel

◁ *Historic nests. House martins may use the same nest for several years, providing it is in good condition. These house martins' nests are built in crevices in a rock wall, one of the old sites for their nests before man provided houses and overhanging ledges.*

▽ *A suburban estate: rows and rows of sand martins' nests. The nests, at the end of a long tunnel, are made of grass, plant stems, and other birds' feathers. Although close together the nests are built completely independently, unlike those of the house martins.*

the young bird, but it bides its time and in due course takes strongly to the wing.

Many young birds leave the nest and spend a short time hopping and fluttering about the branches near the nest. Martins cannot do this; they have to fly strongly almost immediately. Presumably this is why the adult birds show so much concern, but it is difficult to see why some martins fly prematurely. Sometimes they wait only half an hour or so before taking flight again. On occasions, however, young martins have been found on the ground and taken to the shelter of a shed or aviary. There they sit, apparently unconcerned, for as much as a day, then suddenly take off and fly strongly.

It is not known what factors cause a young bird to make its maiden flight. It certainly does not have to practise or learn to fly like we have to learn to swim. Chicks can often be seen flapping their wings vigorously but this seems to be no more than taking exercise. Chicks that are raised in conditions where they cannot flap their wings, as in a cramped martin's nest, can fly just as readily as those allowed to flap their wings. It seems that the ability to fly is instinctive, but the delicate control of flight, including landing, has to be learned.

Fritz Siedel

class	**Aves**
order	**Passeriformes**
family	**Hirundinidae**
genera & species	**Delichon urbica** *house martin* **Petrochelidon nigricans** *tree martin* **Phaeoprogne tapera** *brown-chested martin* **Phedina borbonica** *Mascarene martin* **Progne subis** *purple martin* **Pseudochelidon eurystomina** *river martin* **Ptyonoprogne rupestris** *crag martin* **Riparia riparia** *sand martin*

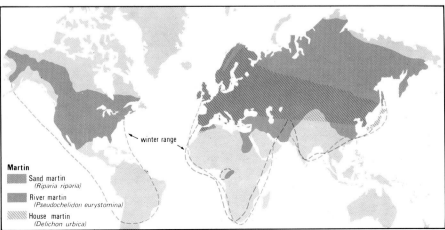

winter range

Martin

Sand martin
(*Riparia riparia*)

River martin
(*Pseudochelidon eurystomina*)

House martin
(*Delichon urbica*)

Mayfly

Mayflies make up one of the most distinct and peculiar of insect orders. They have features which are not found in any other group of insects, and all the species in the order are very much alike.

*The adult has large forewings and its hindwings are small, sometimes absent altogether. Each wing has a fine network of veins, and when the insect is at rest all four wings are held close together over the back in the manner of a butterfly. The legs are small and weak and the tail ends in three, sometimes two, long filaments or cerci. The eyes are large, especially in the males. In some genera, **Cloeon**, for example,*

the land. The compact mass of moving insects attracts the notice of females, which fly into the swarm. Each is at once seized by a male, the pair then leaving the swarm to mate. The males die almost immediately after mating and the females soon after laying their eggs, although they may have spent several years as aquatic nymphs.

The female mayfly always lays her eggs in water, but in some species she drops them from the air as she flies just above the surface. In many species the eggs are provided with fine threads which anchor them to water plants or pebbles. A few give birth to living young which have hatched from eggs retained in the mother's body. In the egg-laying species one female may lay several hundred, or several thousand, eggs.

The nymphs are always aquatic; breathing is supplemented by gills set along each side

the males have two pairs of compound eyes, one pair with small facets on the sides of the head and one pair with large facets on the top. The antennae are reduced to tiny bristles, suggesting that the insect is aware of its surroundings mainly through its sight. The jaws and other mouthparts are vestigial and functionless and adult mayflies never feed. They only gulp down air until the stomach becomes distended like a balloon. This reduces the insect's overall specific gravity and makes the mating flight easier.

About 1 000 species are known, but only those of Europe and North America have been thoroughly studied and there must be many species still undescribed.

Guiding light

The adult life of mayflies is concerned solely with reproduction. They nearly always hatch in great numbers together, and the males gather in dancing swarms over

of the abdomen. Unlike the adults, the nymphs show many differences from species to species. Some are adapted for swimming actively among water plants, others to living on the bottom, burrowing in the mud or clinging to rocks in rapidly flowing water. Nearly all are vegetable feeders, and they take from a year to as much as 4 years to reach full size. A recent discovery about free-swimming mayfly nymphs is that they orientate themselves in the water not by a sense of balance based on gravity, as had been supposed, but by the direction from which the light reaches their eyes. In an aquarium with a glass bottom, lit from below, they swim upside-down.

They mate and die on land

The event in the mayfly's life which sets the Ephemeroptera, as they are collectively called, wholly apart from all other insects is the change from a subimago to an imago. Before this happens, however, the fully grown nymph rises to the surface of the water and floats there, or crawls out onto

Heather Angel

John Goddard

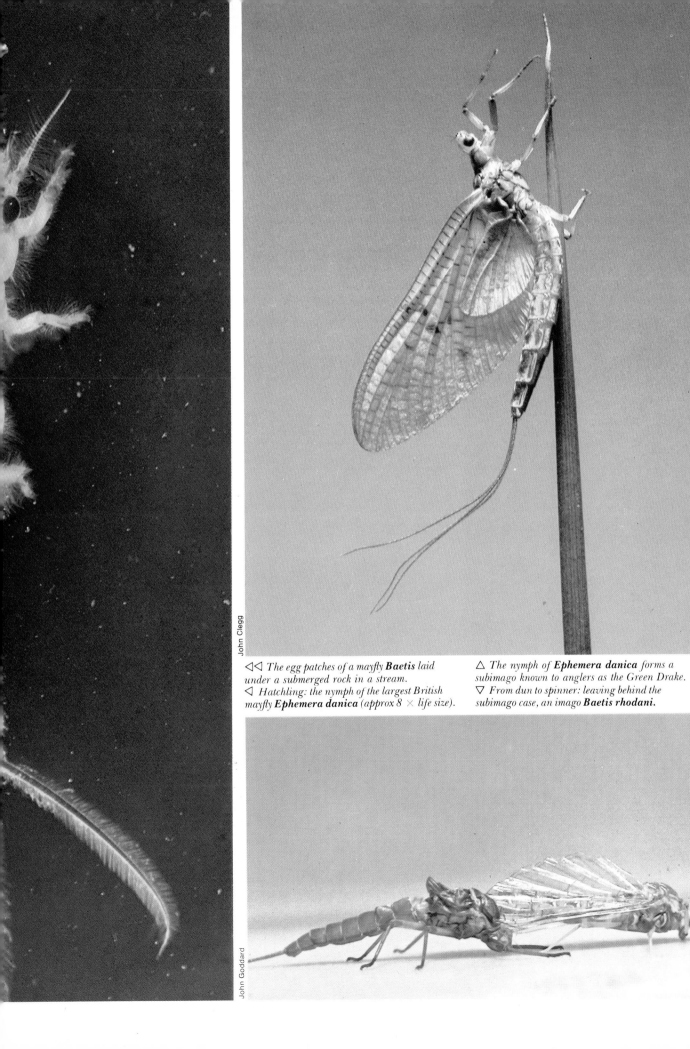

◁◁ *The egg patches of a mayfly* **Baetis** *laid under a submerged rock in a stream.*
◁ *Hatchling: the nymph of the largest British mayfly* **Ephemera danica** *(approx 8 × life size).*

△ *The nymph of* **Ephemera danica** *forms a subimago known to anglers as the Green Drake.*
▽ *From dun to spinner: leaving behind the subimago case, an imago* **Baetis rhodani.**

John Clegg

John Goddard

a stone or reed. Its skin splits and a winged insect creeps out, resting for a time before making a short, rather laboured flight to a bush or fence or to a building near the water. Although it is winged, this insect that creeps laboriously out of the water is not the perfect insect or imago, as the final stage is called. After a few minutes or hours, varying with the species, the subimago moults again, shedding a delicate skin from the whole of its body, legs and even its wings. Then it flies away, much more buoyantly than before, but it will live at most for a few days and, in many species, only for a few hours. A subimago can be recognised by the wings which appear dark and dull due to the presence of microscopic hairs, which form a visible fringe along their hinder edges.

The fisherman's fly

Mayflies are of great interest to anglers, both on account of the part they play in supplying fish with food, and by their direct connection with the sport of fly-fishing. The hatching of a swarm of mayflies excites the fish, especially trout, and stimulates their appetites, so they take a lure more readily than at other times. The swarming is known as a 'hatch' and the fish are said to 'rise' to it. They feed both on the nymphs as they swim to the surface and on the flies, both subimagoes and adults, which touch or fall into the water.

A mayfly is far too delicate an object to be impaled on a hook like a maggot or a worm, and the flyfisherman's practice is to make replicas of the flies by binding carefully prepared scraps of feathers onto the shaft of a hook. These are then 'cast' and allowed to fall onto the surface of the water. A lure so used is called a 'dry fly', as opposed to a 'wet fly', which sinks under the surface.

The more exactly the artificial fly duplicates the species of mayfly which is hatching, the greater are the chances of success, so anglers study mayflies carefully and they have their own names for them. Any subimago is known as a 'dun' and the imagoes are called 'spinners'. Those of the species *Procloeon rufulum* are called the Pale Evening Dun and the Pale Evening Spinner respectively. *Rhithrogena semicolorata* are the Olive Upright Dun and the Olive Upright Spinner. The two stages of *Ephemera danica*, the largest British mayfly, have separate names, the Green Drake and the Spent Gnat.

If a well made artificial fly is swung, by means of a rod and line, into a swarm of the males whose species it represents, numbers of them will pursue it, losing interest only when actual contact reveals to them that they have been heartlessly deceived.

class	**Insecta**
order	**Ephemeroptera**
families	**Ephemeridae** **Caenidae** (hindwings absent) *others*

◁ *A short life but a gay one: having passed through all the hazards of a year or more of aquatic larval life this adult* **Ephemera vulgata** *will live only long enough to reproduce.*

Meadow brown

This is the common name of a butterfly **Maniola jurtina** *of the family Nymphalidae. It is found in areas of waste ground with uncut grass, bushes and hedgerows, and in woodland clearings. Where this type of habitat is allowed to persist it may be very common.*

The sexes are very distinct, the males being smaller and dull coloured, the females larger and brighter with an area of dull orange on the forewings which is much less extensive in the male. Like other butterflies of the family Nymphalidae the male has a patch of scented scales on each forewing, which are important in courtship. Their scent has been described as resembling that of an old cigar box.

The meadow brown ranges over the whole of the British Isles, and over Europe, North Africa and eastward to northern Iran and, in the USSR, as far east as western Siberia. There are local variations in its coloration over the whole of its range and several distinct forms occur in the British Isles.

Unobtrusive caterpillars

In midsummer meadow browns are on the wing, ready for mating. The male approaches the female and the scented scales on his wings, known as androconia, make the female responsive. After mating the female lays whitish green eggs shaped like minute barrels with vertically ribbed surfaces. They are laid on the blades of meadow grass and other grasses, which are the food plant of the larvae. The caterpillar is green with short white hairs and with a darker line along the back and a lighter one on each side. There are two short points or tails at the hind end, a feature almost universal among larvae of the subfamily Satyrinae. The larvae are seldom seen as they feed only at night, hiding at the base of the grass stems by day. They have a long life, from early August to the following April. The green pupa is suspended by the tail from a stem or blade of grass, and the butterfly emerges 3 weeks to a month after pupation.

Races of the meadow brown

The meadow brown butterfly belongs to the subfamily Satyrinae, popularly known as 'browns' from their sombre colouring. However, this was applied by entomologists unfamiliar with the 'browns' of the tropics and subtropics. While the majority fit the name —although often having intricate patterns in brown and orange—there are many brightly coloured tropical species. In South Africa, for example, there are Satyrinae which are blue, and would easily pass as Blue butterflies (Lycaenid). One rather gorgeous satyrid is silver and is most unusual in appearance. In the depths of the

▷△ *A meadow brown caterpillar cuts the ground from under his own feet as it eats away a leaf.*

▷ *Two pupae of the meadow brown, which will soon burst into butterflies, are attached to the plant by a tail at their bases.*

Amazon forest there lives a curious transparent satyrine butterfly; it has been described as being like the delicate petal of a flower floating in the deep shade in which it lives.

Many butterflies alter their position when they land in bright sunshine in order to cast the smallest possible shadow, helping to make themselves less conspicuous. This can be seen in the European grayling butterfly which generally sits on rather exposed ground. On landing here, it can often be seen to lean forward towards the sun's rays, thus reducing its shadow length. In the tropics, some of the satyrine butterflies are out at dusk, unlike the more usual sunloving image they have; these are the evening browns.

Species of the subfamily Satyrinae are worldwide, ranging from the tropics to the Arctic Circle and high up on the mountains. This butterfly has become separated into local populations or races throughout its range probably because it seldom flies far and is rather sluggish in its habits.

Perfumed courtesy

The most complete study of the courtship of butterflies in the subfamily Satyrinae has been of the grayling. This shows the use of the scent patches. The male chases the female through the air until she lands, when the male takes up position facing her and so close that her antennae overlap his head and front part of the thorax. The organs of smell are in the antennae. Once settled into this position the male courts the female, ending in an elegant bow. At the same time he closes his wings which, up to then, have been spread or halfspread. In doing so he catches her antennae between his forewings and they are pressed against the scent patches. Male grayling butterflies that have had the scent patches removed, which can be done without further injury, can court females as ardently as they may, but without the scent organs the chances of getting a mate are extremely small.

phylum	**Arthropoda**
class	**Insecta**
order	**Lepidoptera**
family	**Nymphalidae**
British genera & species	**Aphantopus hyperanthus** *ringlet* **Coenonympha pamphilus** *small heath* **C. tuilia** *large heath* **Erebia aethiops** *Scotch argus* **E. epiphron** *mountain ringlet* **Eumenis semele** *grayling* **Maniola jurtina** *meadow brown* **M. tithonus** *hedge brown or gatekeeper* **Melanargia galathea** *marbled white* **Parage aegeria** *speckled wood* **P. megaera** *wall butterfly*

▷ *Flower power? A meadow brown butterfly takes a drink of nectar before flying off, perhaps to pollinate another flower.*

Stephen Dalton MHPA

Meadow frog

One of the commonest frogs in North America, the meadow frog has a variety of common names, including grass, leopard, or shad frog and 'herring hopper'. Its body is slender, its head pointed, and it measures $2\frac{1}{4}-4$ in. overall. Its colour varies but most commonly the meadow frog is brown or greenish-brown with rows of green, brown or black spots, each ringed with a lighter colour. The legs are bright green and the belly white. Meadow frogs are found from Labrador and Mackenzie in southern Canada through the United States and Mexico to Nicaragua. They are not so common along the Pacific coast as in the central and eastern parts of the United States.

Lives near water

The meadow frog usually lives near streams, lakes or marshes, or even along irrigation streams running through deserts. It is found both up mountains and on plains, its range being limited apparently only by its ability to reach water. Adult frogs may wander a mile or more from open water but the younger ones tend to stay near the banks. In wet places meadow frogs live in crevices or holes to which they return year after year, while in drier areas they make small 'forms' by clearing away a patch of leaf litter. In winter meadow frogs hibernate underwater, in mud or beneath stones.

Varied diet

The food of the meadow frog includes leeches, snails, spiders and many kinds of insects, both terrestrial and aquatic. These include crickets, grasshoppers, houseflies, beetles, backswimmers, caddis flies—even bees and wasps. Meadow frogs also take larger prey such as small fish, tadpoles and smaller frogs, small snakes and, most surprisingly, they have been known to catch small birds such as the ruby-throated hummingbird.

Songs for all occasions

As soon as the meadow frogs come out of hibernation they breed, in April in the more northerly part of their range. The male has three distinct calls or songs. The main song is a low, long, grunting note followed by several short notes and does not carry very far. This is to attract other males and females to the breeding pools. The males indiscriminately grasp any other frog, recog-

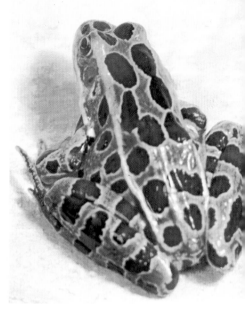

△ *Within the large range covered by the meadow frog there are many colour variations; enough,*

▽ *Portrait of a meadow frog. The bulge made by the top of the pelvis — the long hip bone that makes a platform for frogs' leaping — shows above the head in this unusual view.*

nising them as male or female by touch—the females are swollen with eggs—and by voice. If a male is clasped it utters the warning call to deter the clasping male. The male also calls to tell the female that he intends to clasp her, so preparing her for mating. If she is not ready for mating she warns him off by grunting. The male grasps the female around the shoulders and executes a series of 'backward shuffles', swimming backwards for a short distance then resting.

About 3 000—6 000 eggs are laid in spherical masses of up to 6 in. diameter and fertilised immediately. The masses are usually attached to water plants but may float free in the water. Depending on the water temperature the eggs hatch in 2—3 weeks, liberating tadpoles about ¼ in. long. The tadpoles change into froglets when they are about 1 in. long, after 8—11 weeks' growth. The small frogs live in marshes and begin breeding when 3 years old. A meadow frog lived in the London Zoo for 9 years.

Leaping to safety

Meadow frogs are very good jumpers. They can leap 6 ft, which is 15 times their body length, compared with the bullfrog's jumps of nine times its body length. Most meadow frogs live near enough to water for them to be able to reach the safety of the mud with a few leaps. Experiments have shown that meadow frogs have a simple mechanism that guides them towards water. Their eyes are particularly sensitive to blue light and, if placed in a box with two windows, behind which different coloured screens are stood, a meadow frog jumps towards a blue screen more often than it jumps towards any other colour. By contrast the frogs very rarely jumped towards green.

Putting oneself in the place of a frog in its home in the plants near a lake or stream, all will be green except for the water which reflects blue light. Even on an overcast day when a lake or stream appears grey, the water is still reflecting blue light. When an enemy appears a frog jumps out of the way, but which way is it to jump? The obvious place of safety is in the water where it can hide in the mud or among the weed. The frog's sensitivity to blue light gives it automatic and very rapid guidance.

class	**Amphibia**
order	**Salientia**
family	**Ranidae**
genus & species	*Rana pipiens*

in fact, to make some scientists believe that several separate races are represented.

E Elkan: NHPA

▽ *Leaper extraordinary: for most meadow frogs safety is seldom far away; if disturbed they can reach water in a series of 6ft bounds—a startling performance for their size.*

John Tashjian at Steinhart Aquarium

Menhaden

The menhaden is a member of the herring family which occurs in enormous numbers off the Atlantic coast of the United States, south of Cape Cod. It has a large head with toothless jaws, deep compressed body about 16 in. long and closely set, overlapping bluish silvery scales. It has a single dorsal fin, forked tail and small pectoral and pelvic fins.

The position occupied by the menhaden can best be shown by reference to other members of the herring family. The herring on the eastern Atlantic seaboard goes as far south as the English Channel and south of this it is replaced by the pilchard, as an important food fish. On the western Atlantic seaboard the herring goes as far south as Cape Cod where it is replaced by the menhaden, the alewife and several species of shads.

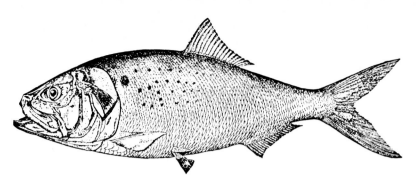

Iain Macmillan

△ *Victimised fish: bluefish decimate menhaden shoals, and creatures as far apart as whales and flounders make meals of them.* ▽ *Man, too, takes his toll: a menhaden boat off New Jersey.*

N Kenneth Ebbs

Whirlpool shoals

The menhaden live in enormous shoals numbering millions of fishes, which feed on plankton in shallow waters and enter estuaries to spawn. Fishes that feed on plankton have a sieve of some kind across the entrance of the throat into the gill-cavities to trap plankton yet allow water to pass across the gills for breathing and out again. The sieve is formed by rods known as gill-rakers. In the menhaden these are unusually long which means it feeds on fine plankton such as diatoms and the smallest animal plankton. One estimate is that a menhaden sieves up to 7 gallons of water in a minute. Dr William Schroeder, the American fisheries expert, writing in 1928, described a shoal of menhaden feeding. They swim swiftly in circles 'like the dust driven by a whirlwind', then suddenly form into a straight line and go forward, rising and falling. They may swim with or against the tide, first in one direction, then in another. As they rise their mouths are wide open, taking in water continuously. When swimming inshore they seldom break surface but farther out they swim in compact shoals just under the surface causing ripples as they rise. At times groups of hundreds leap a few inches out of the water, making a noise audible even 100 yd away.

Fishy millions

Menhaden spawn in estuaries, where they do so according to the temperature of the water. In the Gulf of Mexico spawning takes place in autumn and winter. Farther north spawning is in spring and summer; in the vicinity of New York, for example, it is in May and June. Large numbers of floating eggs are laid and development is similar to that of the herring (p 1196). Authors vie with each other to express vividly the numbers of menhaden in the sea. Dr NB Marshall points out that nearly half the yearly world catch of fish consists of herrings and species closely related to them, that in 1951 the weight of herrings and herring-like fishes caught was 6 400 tons compared with 3 400 tons of cod and

related species and 1 500 tons of mackerel and tuna-like fishes. Dr W Goode, the American author, estimated that the number of menhaden destroyed each year by natural enemies amounts to thousands of times the number taken by man. In fact, the estimates given do not always tally very closely, but at least there is unanimity in suggesting that prodigious numbers of these fishes are caught by man and other enemies annually and this must leave myriads in the sea to carry on the species.

Enemies on all sides

The tally of natural enemies is equally spectacular. The main enemy, according to FD Ommanney, is the bluefish (p 38) which wreaks appalling slaughter among them in summer, sometimes driving shoals of menhaden to the shore 'where they may be seen piled up in rows'. Other destructive predators are whales of various sizes, dolphinfish, tuna and weakfish. In the northern part of their range they are also eaten by flounders. Farther south pompano, cavally, bonito and bayonetfish add to the carnage. To a lesser extent at various points in their range the menhaden are also attacked by striped bass and sea trout. The list could be extended further, merely for those taken by enemies underwater. From the air above seabirds take their toll, gulls especially being attracted to the ripples caused by the rising fish.

No food fish

Large catches of menhaden are made using purse seine nets worked from launches. They are used for fish meal, as fertilisers, for their oil, and as bait. The oil is used in many industrial processes as well as for paint, soap and insecticides.

Spoilt spawning grounds

The herring is essentially a saltwater fish but it occasionally enters freshwater. On the American side of the Atlantic are several related species, in addition to the menhaden, that habitually do this—for instance, the alewife, very like the menhaden but smaller, up to 10 in. long, and the shad, also like the menhaden, but larger, up to 30 in. These two go well up rivers to spawn, and there is the skipjack herring of the Gulf of Mexico that is also widely distributed in the fresh waters of the Mississippi basin. Their enemies include predatory fishes and seabirds, and there are others, like the garpike, waiting for them in the rivers. The biggest menace today for fishes which feed in the sea and ascend rivers to spawn is pollution. Around the British Isles are two species of shad, the allis shad and the twaite shad. These used to spawn in rivers such as the Shannon, Severn and Thames, but today they are of little consequence because of pollution. If river pollution spreads to the seas, as we are warned is likely, the abundant menhaden may even now be under unseen threat in its estuarine spawning grounds.

class	Osteichthyes
order	**Clupeiformes**
family	**Clupeidae**
genera & species	***Alosa alosa*** *allis shad* ***A. finta*** *twaite shad* ***A. sapidissima*** *American shad* ***Brevoortia tyrannus*** *menhaden* ***Pomolobus pseudoharengus*** *alewife*

Merganser

The mergansers, with their very close relatives the goosander and smew, are a group of colourful ducks. They are sometimes called 'saw-bills', because their distinguishing feature is a long, narrow bill set with backward-pointing serrations in both upper and lower mandibles.

The red-breasted merganser breeds in northern latitudes around the world from Siberia, Novaya Zemlya, Scandinavia, Iceland, Greenland, Baffin Island and north Alaska southwards to the Great Lakes of America, Ireland, Scotland and northern Germany. It is about the same size as a mallard but slender in build. The males are easily distinguished by the chestnut breast and the stiff, double crest sticking out from the back of the head like a badly worn paint brush. The head and crest are bottle green and the bill red. There is a white collar around the neck and, apart from the chestnut breast and a white belly, the plumage is black and white. The back is black, there are white bands on the wings and the rest is patterned and scalloped. The female has a brown head and neck and a dull grey-brown body.

The goosander, which has a similar distribution to the hooded merganser, but lives farther south, is slightly larger. The head of the male is bottle green, almost black, with an indistinct mane-like crest. The bill is red, the back mainly black and the neck and underparts white. The female has a chestnut head, light grey back and flanks and white underparts. The smew is much smaller. The male is white with a black patch on the face and a black line on the back. The female has a chestnut head with white cheeks and a grey body. The smew ranges from northern Scandinavia to Kamchatka. Perhaps the most spectacular of the mergansers is the hooded merganser of North America. It is a little larger than the smew and has a fan-like crest, white with a border of black. The rest of the plumage is black and white with chestnut flanks. The female has a crest more like that of a red-breasted merganser.

The remaining mergansers are the rare and shy Brazilian merganser of South America and the Auckland Island merganser. The latter was not at all numerous even when it was first discovered and it is probably now extinct.

Mergansers are migratory, moving south in winter. The smew is a rare winter visitor to the British Isles and the hooded merganser has occasionally crossed the Atlantic to Europe. Outside the breeding season the red-breasted merganser, the smew and the goosander are often found in saltwater although they usually keep to estuaries and inlets. The red-breasted merganser is the most marine of these ducks and may be found on the open sea.

A Christiansen

Roy A Harris & KR Duff

△ The watchful eye. A female goosander watches intently for fish in the clear water.

△ A fish's point of no return. A goosander's bill is adapted for holding its slippery prey.

A Christiansen

1587

Fish-eaters

The backward-pointing serration on the bills of mergansers and their relatives prevents their slippery prey from escaping while being swallowed. Although mainly fish-eaters, they also eat other freshwater animals, such as frogs, worms, beetles and crustaceans. Goosanders have been recorded as feeding on large quantities of trout and salmon but any damage they do to fishing interests is offset by their attacks on predatory fish such as pike and eels which they catch when almost 1 ft long.

The fish-eating habits of mergansers give them a rather unpleasant taste but their fast, direct flight has sometimes made them sought-after targets for wildfowlers. Certainly mergansers are in no danger of extinction and in some places they are increasing in numbers. This is noticeable where new reservoirs form shallow lakes which are good feeding grounds for ducks.

Subaqua mating

The courtship of mergansers is often spectacular, involving much racing to and fro and splashing. Several males may display together before one female. Male goosanders have been seen swimming in front of females, weaving in and out and pushing against one another. Occasionally fighting breaks out and two goosanders rise up in the water, breast-to-breast and throw up sheets of water as they battle together. The males also kick gouts of water 3—4 ft into the air as part of their display. Eventually the female pairs with one of the males and when receptive sinks herself so only a small part of the body can be seen above water. The male mounts her and she completely disappears underwater. Male red-breasted mergansers show off their white ring by stretching their necks and they raise their tails, so they tilt forwards with their breasts submerged and bob about in this peculiar position. At the same time they utter a rough, purring call.

After the mating season the males moult, losing their colourful plumage for the eclipse plumage which resembles that of the females or juveniles. The males take no part in raising the family. The females leave the river or lake where mating took place and build their nests a short distance from the water. Goosanders and smew nest in hollow trees and make use of nesting boxes that in some places are specially put up for them. When there are no trees, as in Iceland, for instance, they nest on the ground, as does the red-breasted merganser, who makes a depression in the ground, protected by rocks or bushes, and lines it with grasses, leaves and down.

The clutch numbers 6—12, sometimes more. When eggs have been taken continuously from under goosanders sitting in nest boxes they have laid up to 40 eggs. Incubation lasts 4—5 weeks according to species and when the chicks hatch out they are led to the water by the parent. When the nest is in a hollow tree, the ducklings jump out and fall to the ground, but they are so light they come to no harm. The ducklings stay with their mother, following her around and sometimes riding on her back, until they are able to find their own food and fend for themselves.

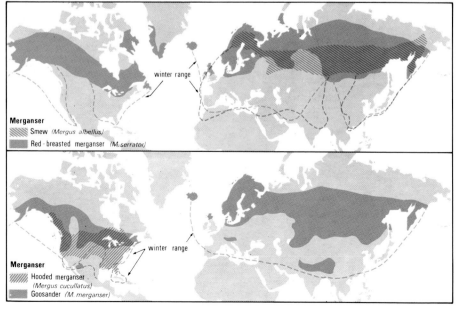

◁ *Overleaf, bottom: Skating party. Goosanders finding, and losing, their feet on the ice.*

△ *Press-ups in the sea: a red-breasted merganser diving for fish.*

G Rüppell

Merganser
////// Smew (*Mergus albellus*)
▓ Red-breasted merganser (*M. serrator*)

winter range

Merganser
////// Hooded merganser (*Mergus cucullatus*)
▓ Goosander (*M. merganser*)

winter range

Ritual movements

A feature of the courtship displays of mergansers is the mock-drinking that also features in the courtship of mallards (p 1520). The duck jabs at the water then stretches its neck upwards just as it does when it is drinking and swallowing. It is thought that this behaviour is now a ritual relic of proper drinking behaviour. During courtship the partners are subjected to three forces—to fight, to flee and to mate. Sometimes there is a conflict between these forces which is expressed by the animal doing something quite out of context, such as drinking, in the same way as when we scratch our heads or whistle when perplexed.

These actions are called displacement activities and in bird courtship they have sometimes developed a function instead of being by-products of a mental state. Drinking movements are now part of the courtship display of ducks and are used as a gesture of peace. In the course of evolution the action has changed slightly and the ducks do not actually drink but just go through the motions. This is called ritualisation; the original action has changed both in form and meaning, as it has among humans. Thus, a salute between men has changed from a gesture to show that no weapon is being carried, to one of greeting.

class	**Aves**
order	**Anseriformes**
family	**Anatidae**
genus & species	***Mergus albellus*** smew
	M. australis
	Auckland Island merganser
	M. cucullatus hooded merganser
	M. merganser goosander
	M. octosetaceus
	Brazilian merganser
	M. serrator red-breasted merganser

▽ *Winter blues: a couple of male smew swimming together in a patch of ice-free water, while a female in the foreground looks on. These ducks are well coloured to camouflage them against their icy background. Their migratory movements in winter are very irregular, like all mergansers, as they depend so much on the local weather conditions. If conditions allow they will spend the winter in ice-free stretches of water within their breeding grounds. On the other hand, they may migrate as far south as the subtropics.*

A Christiansen

Merlin

The merlin, sometimes called the pigeon hawk, is a small falcon. The male, which is smaller than the female and not much larger than a thrush, is slate-blue on the upperparts and reddish-brown with black streaks on the underparts and neck. It lacks the black 'moustache' of the peregrine and hobby. The female is often confused with a female kestrel with its dark brown upperparts and light brown underparts streaked with brown and white.

Merlins breed in both Old and New Worlds. In America they breed north of a line running from California to New-foundland to Arctic Canada and Alaska. In Europe they breed in Iceland, Faeroes, many parts of the British Isles, excluding the southeast, and Scandinavia. From Finland and the Baltic States they spread across northern Asia to Kamchatka. Out-side the breeding season they migrate south usually following the coastline but crossing open water where necessary. The North American merlins fly as far as Ecuador and the British and Scan-dinavian birds move down to the Medi-terranean and North Africa. Icelandic merlins migrate to northeast Britain and some of the Scandinavian population pass through the British Isles.

Bold falcons

When left alone, merlins are bold and quite easy to approach, but when perse-cuted, they become very shy and difficult to observe as they hug the ground, flying behind hedges or hillocks. The numbers of merlins are decreasing in many places, partly through persecution and partly through changes in habitat such as building and reafforestation. They live in open country, breeding in sand dunes, on moors or rough pastures and sometimes on cliffs. They nest in woodland when there is open country nearby where they can hunt. Out-side the breeding season they move into marshland and other low-lying ground. Their flight is buoyant and swift, with rapid wingbeats interspersed with short glides, unlike the hovering of a kestrel or the soar-ing of larger falcons.

Feeds on small birds

Merlins do not use the swift stooping of peregrines and other falcons when hunting but chase their prey in rapid dashes, pounc-ing on it in the air or on the ground. They feed mainly on small birds that live on open moors and pastures, such as larks, pipits, buntings, wheatears and sparrows. Larger birds such as snipe, lapwings, golden plovers and teal are sometimes caught and merlins also catch small mammals including young rabbits, reptiles, dragonflies, moths and beetles. The prey is carried to a perch on the ground, tree or boulder, where it is dismembered.

Ground nesting

Males reach the breeding grounds before the females, arriving in March or April de-

A male merlin calmly eyes the situation. They are bold birds and sometimes quite approachable.

Ronald Austing

pending on latitude. In the northernmost parts of their range they may not arrive until May. Before they pair, the males fly from perch to perch, calling at intervals, and later male and female circle together high over the nest, but merlins do not seem to indulge in the diving and soaring dis-plays of other falcons.

The nesting territory is vigorously de-fended against crows and other birds of prey, even eagles being put to flight. The nest is usually made among low vegetation on the ground, on cliff ledges or among the long grasses in sand dunes. It is little more than a shallow scrape lined with a few strands of vegetation, except in the dunes when quite a considerable structure of marram grass may be made as the sitting bird plucks off stems and adds them to the existing structure. In woodland the aban-doned nests of crows or ravens may be used and on one occasion a merlin nested in a woodpecker's hole.

The eggs are laid with a 2-day interval between each one. There are usually four, sometimes five eggs, but rarely more. Clutches of seven eggs have been recorded in Scandinavia in years when plagues of lemmings (p. 1431) have provided abundant food. Both parents incubate the eggs, the female taking the larger share. During incubation, which lasts 4 weeks, the male brings food for the female, calling her off

the nest and giving her the food either on the ground or passing it to her in the air. When the chicks hatch, the male does most of the hunting, giving the food to the female to take to the chicks. Only when the chicks are nearly full grown does the female assist in hunting. The chicks first fly when a month old. Within 2 weeks they can catch insects and after 6 weeks they can catch birds.

Storing food

Birds of prey often return to any prey that they have not been able to finish at one sitting. This obviously saves the time and trouble of catching more food. The merlin, however, has been seen to store food de-liberately, which recalls the habits of butcherbirds (p.467) and shrikes that impale food on thorns. On several occasions mer-lins have been seen to push the bodies of small birds, pipits or larks, down into clumps of heather so they are hidden from sight, and return later to eat them.

class	Aves
order	Falconiformes
family	Falconidae
genus & species	*Falco columbarius*

*Harem in reverse: the female monia **Monias benschi** mates with several males (foreground) and they build the nests and incubate.*

Mesite

The three birds belonging to the mesite family live in Madagascar. They are oddities difficult to classify and so are usually placed in the order Gruiformes, a rather hotch-potch group that includes cranes, rails, kagus, finfoots and others. One peculiarity is the possession of 5 pairs of powder down patches, more than any other bird and not found in any other of the Gruiformes. At various times it has been suggested that the mesites are related to perching birds, game birds, or pigeons, even that they belong in an order of their own. Mesites are also rather a mystery with regard to their habits, as is often the case with Madagascan animals, simply because, being in a rather remote place, they have not been fully studied.

Mesites are about the size of thrushes with fairly long tails and short wings. The white-breasted mesite is brown above and spotted with black below. It has a grey collar and white on the face and chest, including a very striking white stripe through the eye. The white-breasted mesite lives in the dry forests of the north and west of Madagascar. The brown mesite is rare, living in the rain forests of the eastern side of the island. It is reddish-brown above and chestnut below. The third species is known as Bensch's monia and is quite common in many parts of southwest Madagascar where it lives in low brush forests. The sexes differ in plumage, both are grey above with a white line through the eye and spotting on the chest, but the male has a white throat and chest while the female's throat and chest is a brick red colour.

Unwilling to fly

Mesites live on the ground, running about like pigeons, bobbing their heads and flicking their tails at each step. When disturbed they flee by running, and although some writers say that mesites can fly, AL Rand, the authority on Madagascan birds, was unable to get a captive monia to fly. Even when he threw it into the air it only opened its wings to break the fall as it landed. It is unlikely that they can do more than flutter to the ground after jumping off a branch. Their wings are well-developed but the collarbones are very much reduced. The food of mesites is insects, seeds and fruit which they pick off the ground.

Different duties

There are considerable differences between the breeding habits of the two genera of mesite. In the brown and white-breasted mesites of the genus *Mesoenas* the females appear to do most of the incubation, but the monia, genus *Monias*, appears to be polyandrous, one female mating with several males, the males building the nest and incubating the eggs. Rand has seen groups of monias consisting of one female and several males.

Nesting takes place between October and December, in the rainy season. Nests are built low in the branches of bushes and trees and are reached by the mesites hopping up from one branch to another. Nests of the brown mesite have been found between 3 and 6 ft above the ground. They consisted of platforms of fine interlaced twigs with sides of larger twigs, making a bowl 10 in. across. There is a scanty lining of leaves and moss in which 1−3 eggs are laid. Those of the brown mesite are dull white with a wreath of brown markings at one end. The incubation period is not known but the chicks, clad in brown or black down, leave the nest shortly after hatching.

Isolated lives

The animal life of Madagascar is unusual and mysterious. In the forests that cover much of the island there are many strange animals that have hardly been studied. Unfortunately these forests are now being cut down and many animals, such as the aye-aye (p. 265), are threatened with extinction before we have had time to learn much about them. Apart from being poorly known, Madagascan animals excite curiosity because they are often so different from those of other regions. Although Madagascar is less than 300 miles from Africa at its nearest point its fauna is very different, almost as different as the fauna of Africa is from that of Europe. There are a number of animals on Madagascar that are found nowhere else. There were the flightless elephant birds, now extinct, who laid 2-gallon eggs and nowadays there are three unique families: the mesites, asities and vangas. One strange feature of Madagascan animals is that there has been no proliferation of species as has happened in Darwin's finches (p. 751) and Hawaiian honeycreepers (p. 1225). They seem to be very conservative with no great competition between different animals. There are not many predators on Madagascar which may be why mesites have lost their power of flight, but this seems to be a recent development as they still nest in trees and may, for all we know, still be able to fly a little.

class	**Aves**
order	**Gruiformes**
family	**Mesitornithidae**
genera & species	***Mesoenas unicolor*** brown mesite **M. variegata** white-breasted mesite ***Monias benschi*** Bensch's monia

Midge

Although midges are usually associated with painful bites and irritation, many of them do not bite. There are three families of these small mosquito-like flies that can be properly called midges: the non-biting midges of the family Chironomidae, the biting midges which used to be classed as Chironomidae but are now placed in a separate family, the Ceratopogonidae, and the gall midges, the Cecidomyidae.

There are some 2 000 species of Chironomidae midges, sometimes called harlequin flies. They look very like mosquitoes but are paler and do not suck blood. The biting midges are sometimes barely detectable by the naked eye, but they can produce wounds that itch painfully. They are particularly common in northern regions, and in North America they are called 'no-seeums'. Some species attack

△ **Chironomus** *pupa. The feather 'anal gills' regulate the body fluids which contain red haemoglobin (approx. 10 × life size).*

water, others at the bottom, burrowing in the mud or making tubes of mud and silk in which to live. Chironomid (non-biting midges) larvae have been found at the bottom of Lake Geneva in Switzerland and Lake Superior in North America, and some species live in the sea, at depths of up to 120 ft. The pupae also live at the surface or at the bottom. Those living at the bottom extract enough gases from the water to float to the surface when the adult is due to emerge. The larvae of one of the marine species builds tubes of sand and pieces of seaweed. It lives between high and low water and the adults emerge when the pupae are exposed by the falling tide.

◁ *Emergence: the adult leaving its pupa case.*
▽ *Male midge* **C. plumosus.** *Note the large antennae with their long, sensitive hairs.*

other insects. The gall midges often cause damage to crops; the Hessian fly is a serious pest of wheat in North America and southern Europe.

Also known as midges are the phantom midges **Chaoborus** close relatives of mosquitoes. The phantom or glass larvae are almost transparent with dark, air-filled bladders, one at each end of the body.

Dancing midges

The tiny insects that can be seen on summer evenings dancing up and down in clusters, usually over water, are non-biting midges, performing their mating dance. The males gather in groups and fly up and down; females are attracted to the groups and each mates with one of the males. Most species lay their eggs in water but others lay in rotten wood, dung and decaying plant matter. The larvae are the familiar, small 'worms' that can be seen wriggling energetically in water butts and ponds. Some larvae live near the surface of the

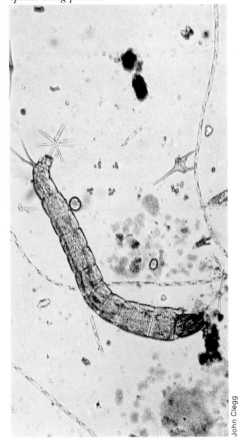

▽ *Larva of **Chironomus** in freshwater plankton; the T-shaped dinoflagellate is **Ceratium,** the star-shaped organism is a primitive plant **Asterionella**. The worm-shaped, plump larva is a very abundant creature in freshwater mud and provides an important food source for fish. In one season in a field study, 2 700 specimens of midge larvae were collected from ½ square metre, some 50 species being present.*

John Clegg

Biting midges

The biting midges are minute, vicious blood-suckers. The adults attack a wide variety of animals, from man to mosquitoes, and feed on their blood, although some feed on nectar. The species that attack mosquitoes suck the blood that the mosquitoes have already extracted from mammals. Other species suck the blood of caterpillars and some pierce the wing veins of adult butter-flies and dragonflies to get their juices. The attacks of biting midges on man can be a serious nuisance, particularly in the evening and in hot weather when the skin is bathed in perspiration. They can penetrate clothing to pierce the skin and, apart from causing severe irritation, some species carry para-sitic worms that cause diseases in man and animals. There are two groups of biting midges. In one the adults have hairy wings and the larvae are terrestrial, living under the bark of trees in manure heaps and similar damp places. The adults of the other group have naked wings and their larvae are aquatic, living in mud or at the surface of the water and swimming by undulations of the body. The minute flies of the family Ceratopogonidae can be easily distinguished from non-biting midges because they carry their wings folded one above the other when they are at rest.

Gall midges

Gall midges are so-called because most of the larvae live in galls on plants. Within the galls, which are formed by the gall midge larvae stimulating the plants to grow ab-normally (see gall wasp, p. 993), the larvae feed on the plant tissues. The larvae of each species of gall midge produce galls of ex-actly the same pattern, a feature which helps identification. The adults of many species also lay their eggs on only certain host plants. Not all gall midges, however, make galls. Some feed on plants without causing galls and others feed on other insects or decaying matter. The most notorious of the gall midges is the Hessian fly, so-called be-cause it is thought to have been introduced to the United States in the straw for the horses of the Hessian troops serving there during the War of Independence. The larvae of the Hessian fly bore into stems of wheat, weakening them, so they fall over. They are extremely difficult to control and cause millions of dollars' worth of damage.

Larval cannibals

A peculiar feature of the life cycles of some gall midges and chironomid midges is that their eggs develop when they are still larvae. This is called paedogenesis and is a form of neoteny (see axolotl, p. 263). The eggs are not fertilised but develop into larvae within the mother's body. They feed on her tissues, killing her and eventually breaking free. The cycle of larvae producing larvae may be repeated many times until overcrowding or a rise in temperature causes them to revert to the normal four-stage cycle.

Unusual bloodworms

The larvae of chironomid midges are red, black or yellow. Red larvae, or bloodworms as they are called, are found at the bottom of ponds where they live in mud. The red colour is due to haemoglobin, the same pig-ment that colours our blood and is used for carrying oxygen. Haemoglobin is rarely found in insects but its presence in chir-onomid larvae is linked with their ability to live in mud where there are only the minutest traces of oxygen. It was originally

◁ *'Big bud' on Juniper. The galls are opened to show the cecid larvae of **Schmidtiella gemmarum**. Since the larvae have such tiny, insignificant mouthparts, it has been a puzzle how they could feed from plant tissues. It has to be assumed that they suck the juices, since they can scarcely be thought to bite and chew.*
▽ *Adult specimen of **S. gemmarum** with eggs.*

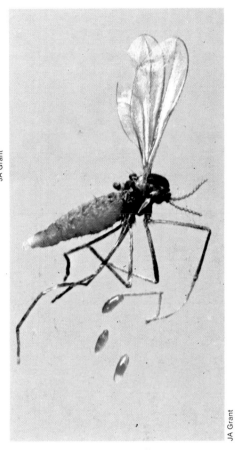

JA Grant

suggested that the haemoglobin stored oxygen for use when the larvae were in poor oxygen conditions, but they have only enough haemoglobin to store oxygen for 12 minutes. It now appears that the haemoglobin acts as an oxygen mop. It has a great affinity for oxygen, combining with it very readily. When a midge larva is in mud with little oxygen in the surrounding water, the haemoglobin sweeps up such oxygen as there is, concentrating it so it can be used in body processes. As a result chironomid midge larvae can live in water, often badly polluted, where no other animals can.

phylum	**Arthropoda**
class	**Insecta**
order	**Diptera**
family	**Chironomidae** *non-biting*
genera	**Chironomus, Tanytarsus,** *others*
family	**Ceratopogonidae** *biting*
genera	**Culicoides, Forcipomyia,** *others*
family	**Cecidomyidae** *gall midges*
genus & species	**Mayetiola destructor** *Hessian fly others*

Klaus Paysan

ME Bacchus

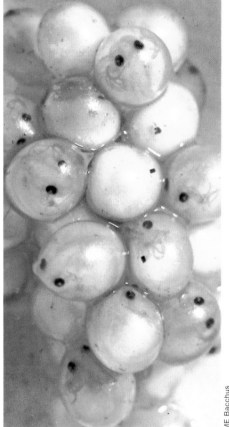

Male midwife carrying the developing eggs. This behaviour is one of the most peculiar types of breeding yet to be seen in amphibians. The male, after fertilising eggs as the female lays them, walks through them so they are wrapped round his hind legs. After about 3 weeks they are ready to hatch and the male enters water so the tadpoles can swim away.
▽ Eggs with visible tadpoles about to hatch.

Midwife toad

The midwife toad is named after its most peculiar mating habits, in which the male apparently helps the female in laying her eggs. Its total length is $1\frac{1}{2} - 2\frac{1}{2}$ in., the female being a little larger than the male. The head is relatively small, the snout pointed and the back is covered with small round warts. The feet are partly webbed, the webs extending only about $\frac{1}{3}$ of the way along the toes. The upperparts are greyish or light brown with a darker patch between the eyes. The underside is dirty white with spots of dark grey on the throat and breast. Females have rows of reddish warts down the sides of the body.

Midwife toads are found in western Europe from Belgium and Germany as far north as Hameln, near Hanover, and southwards to Spain and Portugal, and as far east as Switzerland. They live at heights of 5—6 thousand ft in the Pyrenees where snow lies for most of the year.

Peals of bells

Towards evening these shy, nocturnal toads give away their presence by a whistling call that sounds like a chime of bells when several toads call together. This sound gives them the alternative name of bell toad. Although very abundant in some places, they are seldom seen as they spend the day in holes which they dig with their forelegs and snout, or in crevices between stones and the deserted burrows of small mammals. Midwife toads are quite common in towns, where they hide in cellars or under woodpiles.

When they come out at night midwife toads crawl slowly, rather than hop, in search of insects, slugs and snails. Even when disturbed they can hop only clumsily. They are protected from enemies by the poison in their skin which is secreted from the warts. There is sufficient poison in one midwife toad to kill an adder in a few hours, so any predator that takes a fancy to a midwife toad is unlikely to last long enough to kill another.

Father carries the eggs

The breeding habits of midwife toads are notable for two unusual features; the male stimulates egg-laying by manipulating the female's cloaca, and after the eggs are laid, he carries them wrapped around his hindlegs until they hatch. These habits were not described until 1876 when a Frenchman spent 3 years studying midwife toads.

Breeding starts in April and continues throughout spring and summer. The females are attracted to the males' burrows by their calls. Mating is difficult to watch because it takes place under cover at night, and does not last long. Other frogs and toads stay in amplexus for as much as one day, but in midwife toads amplexus lasts for less than one hour.

The male midwife toad seizes the female around the waist and strokes her cloaca with his hindfeet, even pushing his toes into the cloaca. After a period of up to 20 minutes these movements suddenly stop, the female stretches her hindlegs and the eggs are extruded onto them. The male moves forward, transferring his grip to the females' neck, and fertilises the eggs. At the same time he urinates, soaking the jelly surrounding the eggs, making it swell up.

After fertilisation, the male wraps the string of 20—100 yellowish eggs around his hindlegs and walks off with them. For the next 3 weeks he carries the eggs, resting in his burrow by day and feeding by night. The eggs are kept moist by the dew, but on dry nights he goes to water and immerses them. He may mate again with another female, so carrying two or more strings of eggs. The females also mate again and may produce three or four lots of eggs in one season. When the eggs are ready to hatch, the male enters water and the tadpoles force their way out of the jelly and swim away. How he knows when the eggs are ready to hatch is not known.

The tadpoles are quite advanced when they hatch, being just over $\frac{1}{2}$ in. long and having one gill on each side. Most of them change into frogs before the autumn but some spend the winter as tadpoles and emerge the following spring.

Melting their way out

In early descriptions of the life history of frogs and toads it was assumed that tadpoles bit their way out of the egg capsule. Closer examination revealed that during hatching tadpoles of the midwife toad kept their mouths shut. A hole appears in the egg apparently by the capsule being dissolved away. The whole process of hatching takes about $\frac{1}{2}$ hour. First the tadpole moves so its snout pushes against the capsule which then becomes pointed. After 15—20 minutes the capsule softens, then liquefies completely where the tadpole's snout is pressing against it. Then the tadpole forces its way out.

Between the nostrils of the tadpole is a gland which secretes an enzyme that dissolves the egg capsule. A similar gland had been found in the tadpoles of other frogs and toads whose eggs hatch in water. Those that hatch on land, such as the greenhouse frog (p 1093), bite their way out of the egg.

class	**Amphibia**
order	**Salientia**
family	**Discoglossidae**
genus & species	***Alytes obstetricans***

Migration

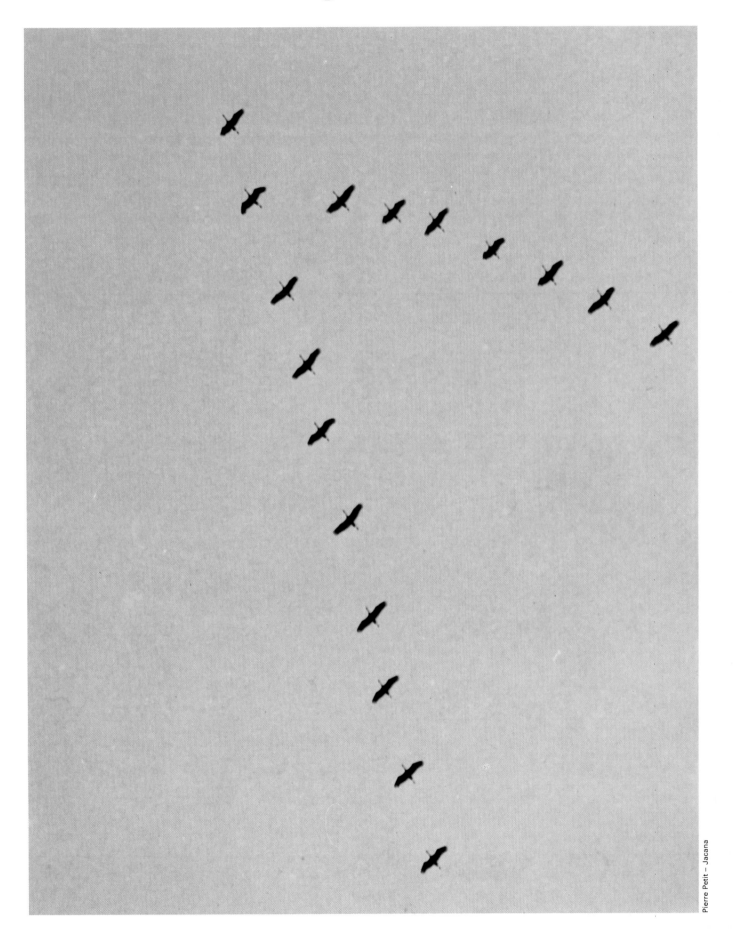

Definition of terms

Migration means a change of location – going from one place to another. But when applied to animals it also implies a subsequent return. This is important because confusion can arise over the undifferentiated use of the word emigration, which, in the sense of leaving an area, corresponds to migration. The difference is, though, that emigration defines the phenomenon of movement in one direction only. In this case, the normal environment is abandoned forever and there is no question of return. Such relocation is also known as acyclic migration. Immigration (the inevitable result of emigration) means a massive resettlement of animals in a new region. Animal invasions or incursions, on the other hand, are irregular in time and place, while nomadism refers to irregular movements within a defined zone.

The principal types of migration

Animal migration, therefore, involves cyclic and periodical relocation and, as well as a return, the phenomenon entails regularity. There are three main types of migration.

Climatic migration is undertaken to escape climatic conditions that are no longer adequate to a species' needs. Alimentary migrations are brought on by two factors, essentially based on food. One is widespread (though temporary) scarcity of food in a given environment. The other is an unusual prolongation of unfavourable meteorological conditions, even though there is still sufficient food. In both cases the species involved will return to the region left when conditions are once again favourable. Genetic migrations are dictated by reproductive needs, and relocation is made to a region favourable for the rearing of young.

Climatic and alimentary migrations, therefore, are directly concerned with the safeguarding of individuals and, as a result, ensuring the survival of the species. Genetic migration, on the other hand, is more directly concerned with the survival of the species.

A great number of species migrate as individuals. However, the concept of mass movement of populations is still valid. For as well as being an established fact, this leads to an understanding of how species fulfil their natural instincts in terms of activity and of the phenomenon of gregariousness. In the same way, alimentary migrations lead to a consideration of the relationships that exist between animals and their environment.

Seasonal migrations

The regular timing of animal migration is chiefly determined by seasons. But, in certain cases, the migratory movements observable today have been influenced by historical factors and originate as the result of previous extensions of territory, or enlargements of an original habitat, towards regions previously unpopulated by the species concerned.

Seasonal migrations reveal the inter-dependence that exists between animal life and all terrestrial and marine biotopes, and the migratory movements of most animals fall into this category. The migratory cycles of most birds and mammals, together with a large number of fish, are annual and directly linked to seasonal cycles. However, there are exceptions. Some species only reproduce once in their lifetime – eels and salmon, for instance – and in such cases the migratory cycles are based upon the lives of individuals, which only return to their birthplace to spawn and die. Similarly, in the case of insects, their short life span means that migrations can only be undertaken by successive generations.

There are three main types of seasonal migration. Some animals migrate from north to south (or the reverse); others migrate from high altitude to lower regions; and others perform strictly local migrations.

The majority of species follow a migratory route that goes from one latitude to another. Animals that migrate from north to south are motivated by unfavourable climatic conditions in the high latitudes neighbouring the poles. They move to escape the winter of one hemisphere, without encountering that of the other.

Altitudinal migration takes place among otherwise non-migratory species (especially in temperate zones) which leave mountains in winter to go down to valleys where they

A diagram of the migration routes used by the Arctic tern. The red areas are the breeding regions; the arrows follow the migration routes.

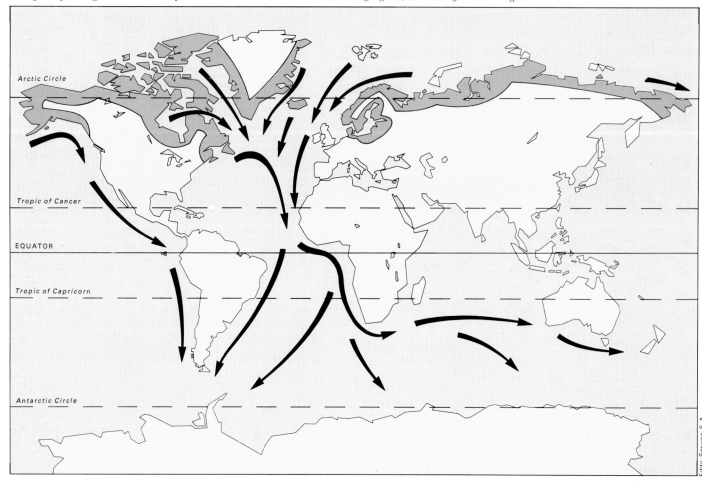

are more likely to find food. This type of vertical movement also takes place in the oceans, where a large number of species of fish and invertebrates, which live in upper waters during the summer, move to deep and warmer water in winter.

Local migrations have a repetitive, if not far reaching, nature. They involve species which live in exposed areas during the summer, but move to seek more sheltered places in which they can spend the winter.

Circadian migrations are another form of restricted migration. These apply to certain species of birds and mammals which spend their rest periods in one place (a 'communal dormitory') and leave it every morning, or evening, to go to another to feed. This done, they then return to their lairs. Bats, crows and rooks are classic examples of this daily cycle, and plankton also has a circadian cycle of vertical migrations.

Migrations vary in direct relation to the species concerned and also local conditions. One tends to think of them as starting from cold countries, but in fact migrations take place from almost every part of the world, including inter-tropical regions.

The migratory impulse

In examining the migratory impulse, one needs to distinguish between movements due to simple taxis, and those properly called migrations, which are a highly complex phenomenon. Taxis is the reaction of an

Jane Burton: Bruce Coleman Ltd.

△ *The steppe lemming* Lagurus lagurus, *like the European lemmings, migrates in large numbers.*

▽ *Arctic terns* Sterna paradisaea macrura. *They fly from the northern coasts of Eurasia and North America right down to the Antarctic.*

Bel & Vienne – Jacana

△ *A herd of reindeer (domesticated caribou* Rangifer tarandus*) in Lapland.*

▽ *In April and May caribou migrate north to feeding areas (light green) in open tundra. In June and July they return south into the coniferous forest (the northern limit of which is shown by the dotted line), but return north again to rutting areas (dark green) in September.*

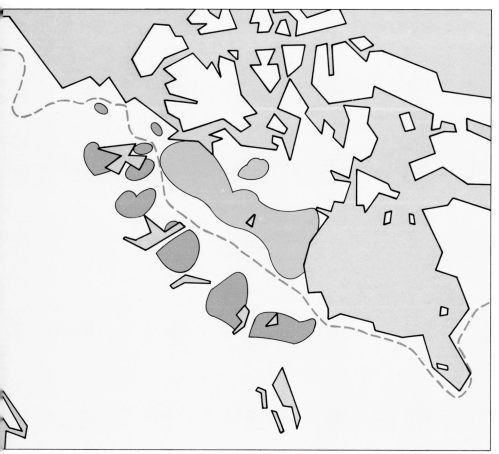

organism to external stimulus by movement. Remove the stimulus and the animal will stay where it is. The migratory impulse is far more complex, involving physiological mechanisms and a variety of factors that bear upon the impulse to move itself.

Migration is integrated in the annual cycle, just as reproduction is. The whole of an organism is involved, including its endocrine glands and gonads, and to really understand why a given species migrates, one needs to examine the metabolism of the animals concerned.

In essence then, migration is a form of behaviour—at one and the same time innate and adaptive—brought on at a specified time by internal and external factors. These factors do not fall into any order of predominance one over another. Nor do the animals concerned decide to migrate—as they are unaware of the conditions to come. They do it instinctively, and some rather better than others, as we shall see.

Animal migration and emigration

Several species of mammals are migratory. However, in general, land mammals make fewer seasonal relocations than birds. Their respective means of moving obviously play a part in this and it is impossible to submit these to any form of comparison. All the same, many of the larger ungulates do cover considerable distances every year—and often according to an unchanging route. Instances of this can be seen in Africa and North America. The North American caribou *Rangifer tarandus* spends the summer in the Arctic tundra and at the end of July goes a good deal further south to the forests where it will winter. The African gnus and zebra also migrate. The South African springbok emigrates and large herds have left game reserves—and never returned.

The common squirrel, in Finland, emigrates towards defined places in movements dictated, it is believed, by a phenomenon of local over-population. The Canadian lynx population undergoes rapid multiplication, in cycles that occur every 10 or 11 years. The surplus animals then go off to colonize territories far to the south. The grey squirrel *Sciurus carolinensis* also emigrates periodically in the United States.

Perhaps the least typical of all animal emigrations is that of lemmings. When a rapid population increase occurs large numbers emigrate and many drown in the sea.

Bats are excellent fliers and though some hibernate, others avoid doing so by undertaking long migrations. This is the case with the North American bats and the boreal bat *Laisurus borealis* which migrate a long way south. It is hard to say quite why certain species of bat migrate and others hibernate; another problem is how bats orientate themselves during movements which naturally take place at night. A great deal of research and observation still needs to be undertaken into the world of bats. All are gregarious and come together in a confined space, or dormitory. So presumably migrators appreciate the possibility of a safe shelter near their initial quarters as much as other bats—but simply do not take advantage of it.

Europe has a number of migratory bats of the genera *Minopterus*, *Pipistrellus* and *Myotis*, which migrate south in winter. The genera

Barbastella and *Tadarida* migrate exactly like birds and sometimes accompany them as they fly through Alpine passes. America has two species which live in Canada in summer – and winter as far south as Florida.

Migratory aquatic mammals

The grey whale *Eschrichtius gibbosus* (p. 1107) is a classic example of a migratory aquatic mammal and moves from its summer quarters in the Bering Sea to Southern California – where it winters and gives birth to its young. This cetacean covers around 6000 miles in 80 to 90 days, following the coastline at a distance of some 2–3 miles from the shore. So aquatic animals can also respond to annual climatic fluctuations by carrying out long migrations, and Antarctic rorquals certainly do. In the course of the southern winter, these baleen whales migrate north, feed round the north-west of Africa, and are also seen in the Gulf of Bengal. The Arctic cetaceans behave in the same way in the Pacific and the Atlantic, with the north Pacific species wintering in the Indian Ocean or the seas around Indonesia.

The Pribilov Islands fur seal also migrates (though it only reproduces on the shores of the Pribilov Islands) and in summer will go as far as Southern California – though the males of the species at this time live near Alaska. It is impossible to know if there are migrating dolphins and porpoises as the area of geographical distribution of their species is so vast. However, herring – the main food of dolphins – do migrate, following the plankton that drifts round the oceans at the mercy of the currents. Thus, it is reasonable to suppose that dolphins also migrate to follow the shoals of herring.

Eels and salmon

The migratory behaviour of fish is not without its mysteries too. Experts have explained the migrations of eels, but, even though it is related to reproduction, the irresistible return of adult eels to the sea must be motivated by a fair amount of blind instinct. Pregnant eels have never been captured and exactly where eels reproduce has long been an enigma. Their reproductive cycle has recently been studied in some detail, however. It begins in the Sargasso Sea, where spawning takes place at a depth of 1600 ft. The adults die and the larvae, called leptocephali, are carried away by the currents. They take a year to reach America and three years to penetrate Europe, and in the course of the journey undergo metamorphosis into elvers. In the rivers the elvers become adult eels, which finally return to the sea. Their morphology and physiology is such that they are able to pass with ease through all the environments they encounter on the way – including the vital transition from fresh to salt water. While they live on the mainland, eels can climb river banks and cross dry land to reach another river or lake.

The salmon is another creature that always returns to its birthplace. In this case, it invariably comes back to spawn in the precise stream where it, itself, was hatched, and to do this, it finds its way across the world.

Fish migration

Sole, herring, tuna and mackerel all mi-

Tollu – Jacana

△ *All species of fur seals belonging to the genus* Arctocephalus *live in the southern hemisphere.*

▽ *The northern fur seal* Callorhinus ursinus, *lives in two separate populations. One group stays in the Pribilof Islands in the summer, migrating down to California. The other stays in the Commander Islands and migrates towards Japan in the winter.*

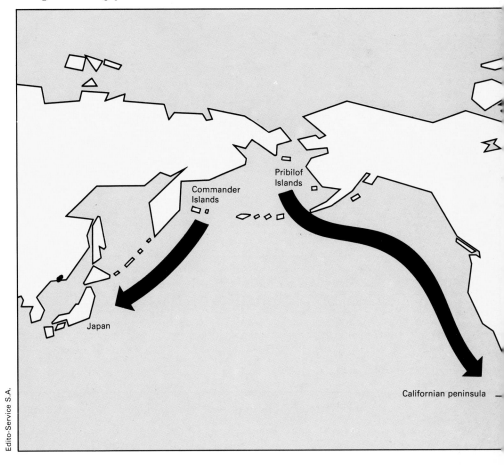

Edito-Service S.A.

Above is the European eel Anguilla anguilla. *Below is a map showing the breeding area (red) and the distribution of the elvers as they increase in size (shown in inches). By the time they reach their freshwater habitats (dark blue) almost all have metamorphosed into adults.*

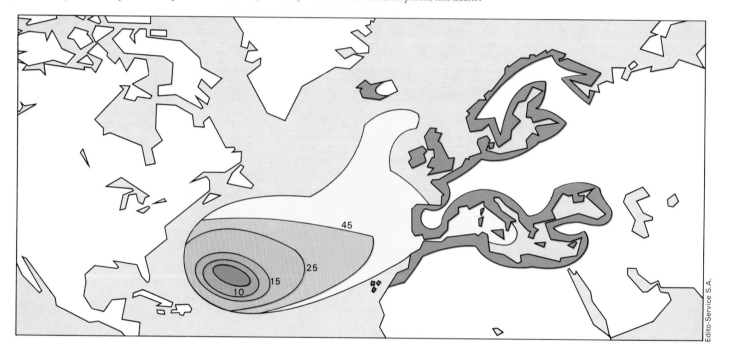

grate–and many other fish may also have seasonal movements which are, as yet, unknown.

The adult mackerel, which is pelagic and migratory, lives in deep water far offshore when it has reached its winter location. Migrating, though, it swims at a rapid rate through sand banks and near the surface. Its geographic distribution extends from Norway to Africa in the eastern Atlantic, and from Labrador to Cape Hatteras on the other side of the Atlantic. Two species of mackerel are seen in the Pacific and it is therefore impossible to speak unilaterally of migrations. Some species are everywhere, which does not help. This is the case with the so-called 'mackerel king' and this example of just one species shows how complex things in the sea can be.

The migration of the bluefin tuna *Thunnus thynnus* would be worth considering in detail. Studied since antiquity, its migratory movements are almost as interesting as those of the eel. But to consider tuna properly, one really needs to enter into history–as it did, penetrating the Mediterranean between the Pillars of Hercules.

Broadly speaking, migratory fish fall into three major categories: species that migrate entirely in the sea; those that migrate in sea and fresh water–like salmon, shad, and sturgeon, which go to spawn in watercourses; and freshwater fish (eel and mullet) that move down rivers and streams to lay their eggs at sea. There is a fourth category of fish that migrate entirely in fresh water, though they are hardly ever studied. It includes trout which, when the time comes to

reproduce, look for especially fast-moving, well-oxygenated water, and to find it will climb high towards the source of their streams.

Amphibians and reptiles

The reign of amphibians may have gone forever, but it is worth remembering that it was a curious, fish-like being which millions of years ago came creeping from the waters to colonize the land. Amphibians not only colonized, but for a time really dominated the entire land surface. Without going into retrospective analysis, it is interesting to consider some of the amphibians that remain.

As any student of nature knows, frogs and toads undertake a form of migration. In-depth studies of the various species of amphibians show that each animal lives in its own defined territory–which it can always find its way back to. Move the animal a good distance away and, if nothing stops it, it will inevitably return to its habitat. On the other hand, all the members of the species (which can be scattered over quite a large area) will come together at fixed times in specific places that were long ago chosen for reproduction.

Toads are drawn to these places as the result of atavism, or a tendency inherited from a remote ancestor. Should a road come between them and their destination, they will be run over sooner than abandon the attempt to reach it. Equally, once such a place has been destroyed, there is no equivalent, as far as toads are concerned–even in neighbouring land that appears to us in every way similar.

For a long time instinct was thought a prohibited term as much by psychologists as biologists, who considered it tainted with metaphysics–and that it attributed to animals (and man) some kind of psychic, as opposed to intelligent, ability. But instinct is really nothing of the sort. The term designates a biological reality, an innate behaviour that need not be limited to trophism or to isolated reflexes. One only has to think of the inexorable journeys of toads to see how true this is.

The spotted salamander *Ambystoma maculatum* is an unusual salamander in that it migrates in groups to spawn. In spring they move at night into ponds, where they pair and spawn. After spawning the females leave the water and do not return. These salamanders use the same paths year after year.

Though the migratory behaviour of reptiles is not as marked as that of amphibians, they also relocate themselves when the time comes for reproduction. Their ways may be solitary, but nature makes its demands nonetheless.

Worms and crustaceans

There are many kinds of polychaete (marine segmented worms) but the palolo worm behaves quite distinctively in its reproductive cycle–which, first and foremost, is tied to phases of the moon. One needs to distinguish between the Samoan palolo worm *Eunice viridis*, which lives in coral reefs

Many fish undertake vertical migrations in the sea, following the movements of the plankton. The plankton itself contains many fish larvae. Those of flatfish, such as this sole, migrate downwards and the adults remain permanently on the bottom.

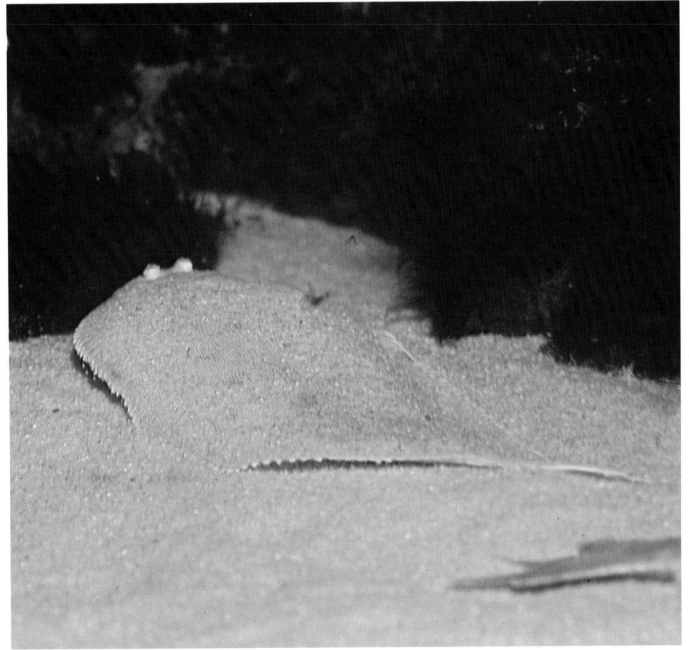

Jack Dermid: Bruce Coleman Ltd.

△ *The spotted salamander* Ambystoma maculatum, *migrates in groups to spawn.*

▷ *The edible frog spawns in shallow water at the edges of pools.*

▽ *A green turtle returning to the sea after laying her eggs.*

Alain Compost: Bruce Coleman Ltd.

1602

The painted lady, or thistle butterfly, Vanessa cardui *migrates northwards, but does not return.*

C. et. M. Moiton – Jacana

around Samoa and the Fiji Islands, and the Atlantic palolo worm *E. fucata* of the reefs of Bermuda and the West Indies. But in warm seas, the palolo worm has an annual cycle of rare precision – it only reproduces at the time of the last lunar quarter, and the natives of its regions have a holiday at this time, when they go fishing for it.

The worm only produces eggs and sperm in the lower part of its body. This is attracted to light, and will float on the surface of the water, independent of the rest of the worm's body. There are so many palolos that these segments of worm form a thick layer on the water – which the natives scoop up in baskets.

Marine crabs may move relatively slowly, but they go a very long way. The females come to spawn along the coasts and then return to deep water more than 120 miles away. The males are non-migratory and stay where they are.

Freshwater crabs move into brackish or salt water when the time comes to reproduce. The Chinese crab *Eriocheir sinensis* lives for some years in fresh water, then goes to salt water to reproduce – some miles from the shore. The young stay in the sea for one year then head towards rivers for as long as a five-year period. Certain tropical crabs that have become terrestrial, instinctively remember a far-distant past and migrate towards the sea when the time comes to spawn.

Insect migrations

The term 'migratory insects' tends to be used too broadly. Many insects alter their location, but they do not all migrate, at least not in the strictest sense of the word. For the movement of an insect population, even on a relatively massive scale, cannot be considered a migration in view of the brief span of their lives, which do not extend beyond a single season. Insects hatch out, leave their

birthplace, spawn in their turn elsewhere, and die. The case of migratory locusts illustrates this kind of emigration as well as anything can. Their movements are referred to as an invasion – and are exactly that.

While mayflies have an ultra-short life span, that of other insects, while only perceptibly longer, does sometimes give them time to relocate themselves. But how these relocations are controlled is beyond our understanding. As we have said, insect movement may involve departure and return – but it is not the same generation of insects that finally makes it back. Succeeding generations then, return to the starting point of their ancestors. But how this is done and the ancestral habitat recognized is not yet really known. Heredity, in the sense of genetically transmitted information, is presumably involved, but the exact mechanism remains unexplained.

There is a surprisingly limited number of

Migrating monarch butterflies Danaeus plexippus *gathering before moving on.*

Roy Pinney – Photo Lib. Inc.

serious studies on the subject of insect behaviour–yet insects are the most numerous creatures on Earth. We talk of invasions of grasshoppers and have always done so, but the reasons for them remain a mystery. They are most likely the result of a population-regulating mechanism within the species, which remains no less a blight all the same. In passing, it is interesting to note that of the seven known species of the grasshopper family, not all display the same propensity to mass relocation–and those that do, do not accomplish this in the same manner.

Diurnal butterflies manage extensive relocations. It is hard to know which of the 250 species to mention to illustrate the extent of their movement. Two European butterflies that set out for the north in summer are the *Colias croceus* and *Vanessa cardui*.

The maybug *Melolontha melolontha* leaves the area where it reproduces to go and feed elsewhere, and other related insects of this type all leave their spawning area to look for food in another–where, what is more, a change in their gonads will take place. They then return to the original area to spawn. Mosquitoes go off to look for blood which helps the females bring their eggs to maturity. Even if all these movements are limited, they are a sufficiently regular phenomenon, for one to liken them to a migration.

The North American ladybird *Hippodamia convergens* goes regularly, in autumn, from the bottoms of valleys to a higher region where it winters. Millions of these ladybirds emerge from under rocks and move together.

The monarch butterfly *Danaus plexippus*, which is an American species and the best known of all diurnal migratory butterflies, has a generational pattern of migration. The number of generations they produce per year varies in direct relation to their habitat: five in the south, one in the north. Speaking of it as the monarch, in the singular, is somewhat misleading, for the butterfly that returns is not the same one that left. But it is the only species of butterfly to journey in this way, in a single and same direction, each year–returning by the same route to its departure point the following year. It always rests at night and in the same places. The flight northwards towards Canada occurs in spring; the return journey, in autumn, as far as Florida, where the butterfly winters. All told, this represents a voyage of 1800 miles.

After hibernating during the winter, the

The northern monarch butterfly (red) migrates southwards from its winter feeding grounds (pale red) and is even found in Australasia. The southern monarch migrates northwards from its winter feeding grounds (pale yellow).

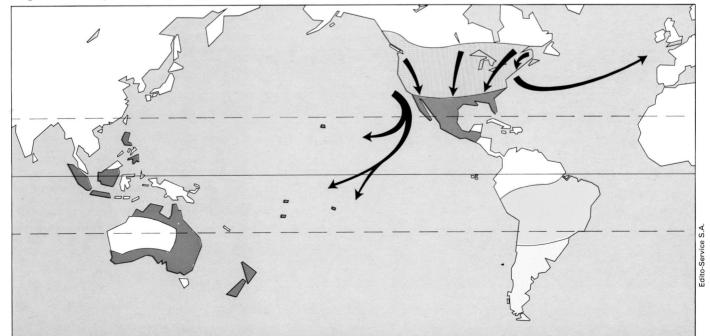

Edito-Service S.A.

Roth – Okapia

△ *A male red-backed shrike feeding his young.*

▽ *The migration route of the red-backed shrike. The main breeding area is shown enclosed by the red line, but a number also breed in Asia even as far as China. The European migrators congregate in the autumn around Greece, leaving there for the south of Africa, where they spend the winter. They return in spring along approximately the same route.*

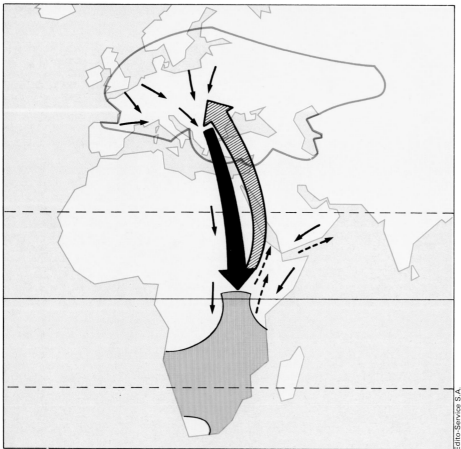

Edito-Service S.A.

monarchs set out again in spring for the north, and the northern parts of their habitat are colonized by successive generations. This means both sexes of butterfly leave and the females lay eggs along the way. In the course of the summer, two or three generations succeed each other–and the new butterflies of each generation continue to fly north. Only in June, at the earliest, do some of them reach Canada–and those that then arrive are perhaps the great-grandchildren of the butterflies that left the region in the autumn of the previous year. The monarch which makes this journey only does it therefore as an entire species, which is even more of a marvel. There is another monarch which performs the same north-south movement in South America, and other monarchs are able to cross the Pacific and reach Australia, as well as other places on the way, such as Hawaii.

Bird migrations

Birds probably have the most complex migratory behaviour and they oblige one to consider longer distances and biological questions. For, due to their high metabolism and need for food, birds are largely dependent on surrounding conditions. These small vertebrates, in fact, are astonishing, and bird migration is one of the most extraordinary phenomena of the animal world. Migration occurs on a wide front and on a massive continental scale. Billions of birds are involved.

The migration of temperate climate birds

The gulls and terns of the family Laridae undertake journeys at regular periods towards the most varied horizons. Dispersal after mating should not be confused with migration; nor should the unusual occupation or invasion of a territory. The northern waxwing, pine crossbill and nutcracker can abandon their northern forests, with no intention of returning, to go south or west. Equally, species can progressively extend their territory and become non-migratory, as the turtle dove has done in western Europe. So one also needs to distinguish between momentary invasion and long-term occupation through progressive acclimatization.

In general European birds adapt well to seasonal fluctuations of climate and food supplies, and the species of western Europe are well off in comparison with northern or eastern species which are obliged to be more migratory. In fact, a good proportion of the birds of all the more temperate regions of western Europe winter where they are. But these regions also have 'visitors', which come to take up winter quarters from the eastern or northern parts of the continent, where they normally nest. Some birds of temperate regions also travel to tropical Africa, which is the predominant winter residence for the migratory birds of the western Palaearctic region.

There are two great axes of migration for Europe, one orientated north–east/south–west; the other north–west/south–east.

▷ *Mallard taking off. They migrate south for the winter.*

Devez-CNRS – Jacana

Cyril Lambscher: Bruce Coleman Ltd.

There are 175 migratory species of birds out of Europe's overall total of around 450 and partial migrators account for another 60 species. Of the 175 true migratory species, 150 take the first north–east/south–west axis; but only 35 go all the way to the end of the journey–to tropical Africa. Others stop before that at various stages on the way. Similarly, 25 species take the alternative north–west/south–east axis, but not all go all the way to the Orient.

Independently of what sort of journey the species make (a decision inscribed in their genetic make-up), enormous losses can occur on the way as a matter of course. Many

△ *Swallows assemble before migrating.*

◁ *A white-throated swallow.*

▽ *Northern migratory birds fly south in the autumn, keeping over land as much as possible.*

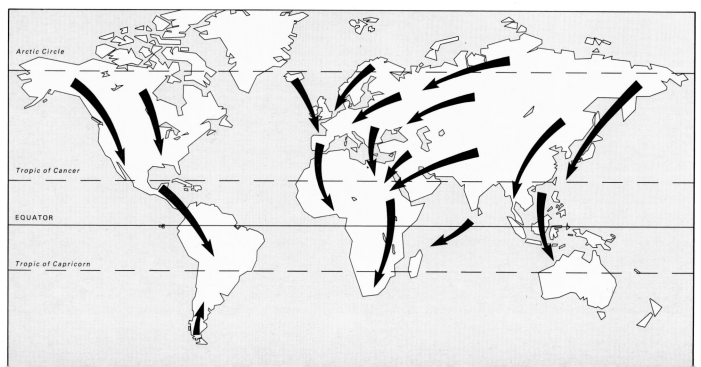

Arctic Circle

Tropic of Cancer

EQUATOR

Tropic of Capricorn

Edito-Service S.A.

F. Winner – Jacana

individuals that set out for a destination lack the ability to reach it. What these destinations are, how birds behave when they get there, when and how they prepare their return, and if they follow the same flight-paths back, are questions that will be considered in a moment.

Precisely which birds leave when? Do they journey by day or by night? Do individuals of the same species, that have come from different places in Europe, all go together at a given moment towards their far-off destination? And to what extent do they winter in different geographical areas? Migrations seem to be well regulated but a number of

L. R. Dawson: Bruce Coleman Ltd.

△ *Black-headed gulls* Larus ridibundus.

▷ *A female ruddy turnstone* Arenia interpres.

▽ *The black-headed gull's summer breeding range (solid red) and winter range (dotted line).*

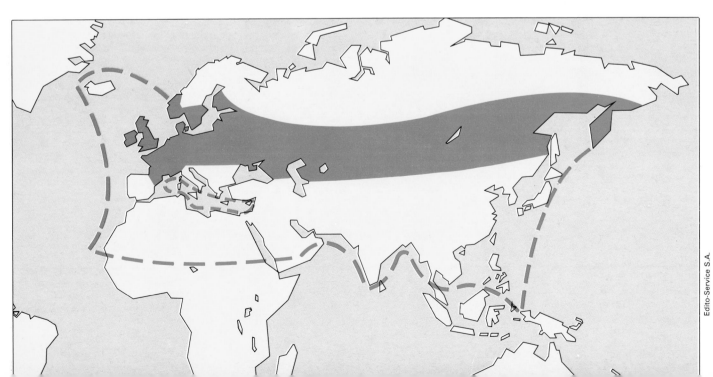

Edito-Service S.A.

Migrating white storks Ciconia ciconia *make their way to Africa either across the Bosphorus or across the Straits of Gibraltar. A few white storks stay in South Africa and Zimbabwe to breed.*

facts still need to be established. We do know, however, that the European migrators destined to go the furthest south will avoid the great equatorial forest, which might seem very tempting after crossing the Mediterranean and Sahara, and that the sparrow family establish themselves in more or less wooded savanna around this great forest, but not especially inside it.

If one examines migratory paths, one sees that some birds are reluctant to fly directly across the sea—storks and birds of prey, for instance, which migrate by day, and the actual way in which they fly plays a part in this. Other migrators travel by night and fly more boldly over the sea, taking direct lines which reduce the journey.

Only the complete case histories of chosen species can fully reveal the astonishing phenomenon of migration. At the present time, not all migratory paths are precisely known. Still more ornithological expeditions, extended use of radar, and consistent tagging are needed to fill the gaps in our knowledge.

Non-migrators are chiefly found among vegetarian species and primarily the granivores. Most omnivores are well adapted to their chosen ecological environment and acclimatized to the conditions of present-day life in Europe. In other words, those birds that can find food in winter in their nesting areas stay there. Insectivores have problems, though, as insects can be scarce or even non-existent in winter. So these birds must leave their nesting area, even if it lies in a temperate zone, and go to warmer regions. There are, of course, some exceptions; the Mediterranean region provides a perfectly suitable place to live all year round—and the black-headed warbler, for instance, takes advantage of this.

The family Laridae (gulls and terns) can be considered here as a 'land' group, because they are coastal and not deep-sea birds and remain on the verge of the mainland. The laughing gull, in fact, lives almost as much in the interior as it does in the exclusively shoreline region and winters fairly far to the centre of Europe. Summer visitors to western Europe find the winter too harsh for them, though species which come from northern regions thrive in it. Sanderling's sandpiper and the collared turnstone appear on European shores from June on, and others will come if conditions become too cold in the north.

One astonishing facet of migration is the gregariousness it can bring about. A pair of sparrows, for instance, that have lived fiercely alone throughout their reproductive

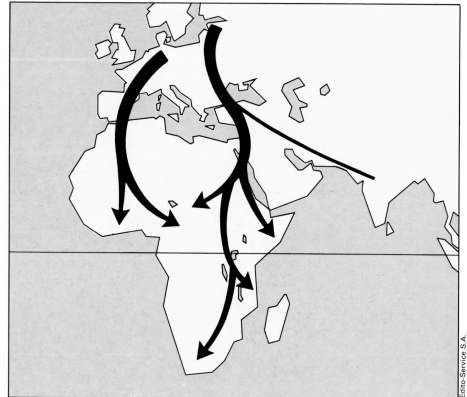

◁As with migratory birds of prey, white storks avoid long sea crossings. They soar high over the land and then glide across the sea. The two streams pass through East and West Africa and the eastern stream ends up in south-east Africa. The populations that breed farthest east in Europe make their way to southern Asia.

△ *Shearwaters are relatives of the albatrosses and fulmars. The Manx shearwater* Puffinus puffinus *migrates as far as South America and the great shearwater* Puffinus gravis *breeds in Tristan da Cunha and migrates to Europe.*

Marc Waren - Jacana

period, will quite happily associate with other birds when the time comes for departure. This is probably dictated by necessity, and dispersal will occur again on arrival. On the other hand, the case of herons and egrets forms an interesting contrast. They tend to nest in colonies, but migrate singly or in pairs.

The birds of North America

Birds that nest in North America encounter similar winter conditions to those that nest in Europe, and the tendency to migrate is especially manifest in the northern continent, as its species often originated in tropical regions. The mechanism of migration is the same: north-to-south, returning by the same flight-paths, which are essentially over the mainland. In this respect, American migrators differ from European birds as they are not obliged to cross sea and great deserts. The Gulf of Mexico can be avoided, even if some species do, in fact, cross it.

The migrations of tropical birds

In general, European birds avoid the great African equatorial forest. (Their wintering areas need further study, for there are invisible lines of demarcation that divide the flood of migrators.) However, there are many species peculiar to the equatorial rain forest of Africa. And what occurs in Africa can be applied, in general terms, to similar latitudes on other continents. Equatorial forest birds have effectively no need whatsoever to move, and though they can be erratic, they are not

migrators. In contrast, species nesting on the savanna, in view of the alternation of dry and rainy seasons there, travel considerable distances. Again, a predominance of insectivores and frugivores among the species of tropical countries make migrations necessary, and here too granivores are favoured.

The Abdim stork nests from the Red Sea to Senegal during the rainy season and winters in southern Africa. But certain birds never cross the equator. Nesting near rain forest, they migrate north in the rainy season. In general, being able to cover great distances, many birds live in perpetual spring.

The migrations of sea birds

As opposed to birds that live beside the sea, those birds that live actually on the high seas, are termed pelagic.

The Bassan gannet (a coastal bird) journeys from Great Britain as far as Senegal. The black-headed gull travels along the coastlines of the Americas, on both the

▷ *The short-tailed shearwater* Puffinus tenuirostris *breeds in south-eastern Australia. Outside the breeding season it migrates all round the Pacific Ocean, making use of the prevailing winds (red arrows). The timing of its 20 000-mile journey is so precise that it ends up in Australia for the same 11 days each year.*

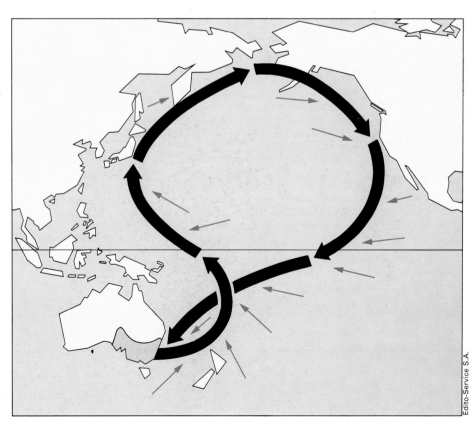

Edito-Service S.A.

Pink-footed geese Anser fabalis *flying by moonlight. They live in Greenland, Iceland and Spitzbergen, and in winter migrate to Britain, northern France, Belgium, Holland and Germany.*

eastern and western sides of the two continents. Reproducing either on the shoreline around California, or on Atlantic shores in the northern United States, the Californian nester then migrates to Peru, while the Atlantic nester flies to Brazil.

The southern continents of Latin America and Africa have a good many migratory coastal birds. Along the Pacific coast, they go north, taking advantage of food carried by the Humboldt current. While in Africa, during the dry season, the Buenguela current feeds Cape gannets as far north as Gabon, as well as terns which in autumn return south to nest. Some coastal birds—such as three-toed gulls, skuas, and arctic terns—also undertake deep-sea crossings.

The pelagic, or deep-sea, birds are the greatest migrators of all. They are chiefly members of the order Procellariiformes: puffins, petrels, and albatross. Some journey throughout the oceans. The Antarctic albatross circumnavigates the world.

Penguins, which at certain times of the year lead an essentially deep-sea life, also undertake extensive migrations; and, as they cannot fly, they do so by swimming. There are also certain land birds for which the sea holds no perils. Some make journeys that cross the North Atlantic, the Mediterranean, or the Gulf of Mexico, and some migrators fly from India to Madagascar. The New Zealand bronze cuckoo winters in the Bismarck archipelago. The long-tailed New Zealand cuckoo migrates as far as the most remote South Sea Islands—a journey of thousands of miles, and the endurance of some of these land birds almost matches that of deep-sea ones.

Atmospheric conditions have an important influence on migrations, and birds can modify their routes according to wind direction and velocity. The height at which they fly varies greatly according to species. The sparrow family flies low, at heights of about 200 ft. Some geese, on the other hand, fly as high as 10 000 ft, and the Indian goose holds an extraordinary altitude record. It has been seen at the incredible height of 30 000 ft.

In general, the altitude at which migrators fly is governed by prevailing conditions and local air currents; comparative tables of various species' altitudes are only of value in relation to specific regions. The speed some birds can maintain, though, is staggering. The bald-headed shrike, for instance, travels at night, between dusk and dawn, for some ten hours—and, in this space of time, can cover a distance of some 300 miles, though, in fact, it takes five days to cover 600 miles. During this time, only two nights are spent flying. The remaining three nights are spent resting and the daylight hours are occupied in feeding to obtain the necessary energy to complete the journey.

Pelicans, storks, birds of prey, swifts and swallows all migrate by day—limicolae, cuckoos, and insectivorous perching birds by night. The flights occur in massed flocks, or in well-ordered formations. The famous migratory 'V' forms an excellent victory symbol—for the triumph of natural selection.

△ *Cracking the eggs. Millipedes hatching.*
▷ *Changing outline. A millipede of the*
order Spirobolida nibbles at an orange leaf.
▽ *A pill millipede, Sphaerotherium.*

WD Haacke

GS Giacomelli

Anthony Bannister: NHPA

Millipede

Millipedes range from being soft-bodied and
less than 1/10 in. long, to heavily armoured
forms exceeding 8 in. Many are poisonous.
One of the most obvious ways in which they
differ from centipedes, with which they used
to be classified under the name 'Myriapoda',
is in having two pairs of legs on most of
their body segments—the first 4 segments
have only single pairs—and in the variable
number of segments. The head has short
antennae, and some species have no eyes.
The general body surface may, however,
be sensitive to light.

Millipedes are typically light-shy,
nocturnal animals living in moist soil,
leaf mould and crevices. They feed on
vegetable matter and though valuable in
breaking this down, some cause much
damage by attacking crops especially in
wet weather. The 1/2 in. spotted snake
millipede Blaniulus guttulatus, for
instance, eats potatoes. The potatoes are
probably attacked only when the tough
skin has already been broken, for the jaws
of these and other millipedes are not
powerful. They then burrow inside, making
them unfit for human consumption.

Chemical warfare

In contrast with the fast-running centipedes,
millipedes are built not for speed but for
pushing their way powerfully through soil or
vegetation. When walking, each leg is a little
out of step with the one in front, so waves
appear to sweep back along each side of the
body. When attacked, some make a rapid
escape without using their legs, writhing and
wriggling their bodies through the vegetation, while others have a protective reflex of
coiling up like a watch spring as soon as they
are disturbed. The pill-millipedes curl into
small balls, and in warmer parts of the world
there are millipedes that roll into golf ball
size. The usual defence of millipedes is,
however, a row of poison glands down each
side of the body. The secretion may be red,
yellow, white or clear, but typically it is
brown or yellow-brown and smelling of
excrement, chlorine or prussic acid. The
latter two are probably present, as well as
iodine or quinine, so it is not surprising that
some species are unpleasant to handle. Indeed, in central Mexico one species is ground
up with various plants as an arrow poison.
Usually the venom simply oozes from the
glands, but some larger tropical millipedes
discharge it as a fine spray, even to distances
of up to a yard. Such millipedes may blind
the incautious chicken molesting it, or it
may discharge onto human skin, making it
blacken and peel. Some of the large tropical

millipedes sport contrasting, bright warning
colours. There are even a few luminous
species. One, sometimes conspicuous at
night in the Sequoia forests of California and
aptly named *Luminodesmus sequoiae*, is blind
and its light shines continuously from
the time of hatching. The light is probably
a warning signal to potential predators
rather than a recognition sign for its fellow
millipedes. That the warning is no mere
bluff was shown by the death, from cyanide
poisoning, of a bee accidentally imprisoned
with one of these millipedes in a test tube.

Tents and mud huts

In mating, which may last for several hours,
the male embraces the female, lower surface
to lower surface. The genital openings are in
pairs on the third body segment and fertilisation takes place inside the female. According to species, 10−300 eggs may be laid.
Some millipedes simply coat each egg with
soil and excrement, so disguising it, and
leave it in a crevice in the earth, while the
females of other species make elaborate
nests and may remain coiled tightly around
them for a few days. The nest can take various forms: a hollow sphere of soil and
saliva, lined with excrement or a thin-walled
dome of excrement, anchored to the sub-
stratum with a narrow tubular chimney, and
covered with bits of leaf. The mother will
replace these if they are removed. Some-

S Bisserôt

times the nest is a tent spun from silk. Some millipedes conceal themselves in such silken chambers or tents when they moult, for having cast their skins, they are temporarily very vulnerable to attack. Usually a millipede eats its cast skin, and also the silk tent. The young millipede starts life without the full number of legs, perhaps only three pairs, and acquires more at successive moults.

Millipedes by the million

In 1878, a train was brought to a halt in Hungary by a black mass of millipedes that carpeted the ground and made the wheels slip on the rails. Trains were again stopped in this way in northern France in 1900 and mass migrations of millipedes of various species have been recorded periodically in various countries, including once in Britain,

in 1885, when a large number were seen crossing a road. More than a score of such phenomena have been recorded in the United States as, for instance, when 75 acres of West Virginia farmland were covered by these animals in 1918. Cattle would not graze and men hoeing in the fields became nauseated and dizzy from the smell. In the end, most of the millipedes, about 65 million, were killed when they were halted at a cliff bottom and were parched by the sun.

Such plagues are rare, but sudden attacks on crops in lesser numbers are familiar, and tend to occur in times of drought following damp weather. The size of millipede populations does not seem to be governed by predators and parasites, though there are plenty of these, including spiders, toads and birds, particularly starlings. The main controlling factors seem to be physical condi-

tions. The best conditions for the buildup of a population occur when there is plenty of moisture and organic matter in the soil, as for example when farmyard manure has been spread. If the soil then becomes dry, the millipedes move to a more congenial environment. What then could be better than a damp cavity in a sugar beet? Sometimes the migrating masses of millipedes have been accompanied by centipedes and woodlice, for reasons far from clear. Conversely, it is not unknown for millipedes to accompany the marching columns of army ants. Again it is not known why, but millipedes are among the 'guests' to be found in the nests of ants and termites.

| phylum | **Arthropoda** |
| class | **Diplopoda** |

▽ *A long embrace. Millipedes, entwined around each other head to head and lower surface to lower surface, may take several hours over mating. These millipedes belong to the order Spirostreptida.*

△ **Cylindroiulus.** *Walking is a serious business with so many legs — two pairs to a segment.*

Anthony Bannister: NHPA

Minivet

There are 10 species of minivets, small, slender birds related to cuckoo-shrikes (p. 725), yet recognizably different from them, the most constant difference being in the tail. In minivets this is graduated and sharply pointed because the outer feathers are shorter than those in the middle. Nobody seems to know how the name came into being, but it dates from as recently as 1862.

One of the larger minivets has been variously named the flamed, scarlet or orange minivet. It is 8 in. long, the male is black and scarlet above and scarlet below except for its black throat; the female is similarly marked but with black and bright yellow. Young males are coloured like the female. Both sexes of the mountain minivet look like this, too, and it is impossible to tell the two species apart, although the first is a lowland bird and the second a mountain bird so their ranges do not overlap. The fiery minivet is slightly smaller but very like the other two species except that the female has scarlet on the lower back. The ashy minivet is less showy. Its upper parts are grey, the head black, the wings brown and the underparts white, the female being paler than the male and lacking the broad white band he has on his forehead.

Minivets are found in southern and eastern Asia, from Afghanistan to Japan, and south to Malaya and the Philippines.

Jane Burton: Photo Res

Chatterers in the tree-tops

But for their gay and striking colours and melodious chattering, minivets would be relatively unknown because they are shy and keep mainly to the tops of trees in the green jungles. They sometimes come to the ground, on paths through the forests, or into gardens where the jungle comes almost to the garden edge. More commonly they work their way through the treetops, in parties of up to 40, attracting attention by their frequently repeated, lively calls. Each species has its particular habitat: the mountain minivet lives at about 8 000 ft, the flamed minivet of India, China, Malaysia and the Philippines lives from sea level to 8 000 ft and the fiery minivet lives in coastal areas especially where there are casuarina trees. The ashy minivet spends the summer in eastern Asia, including Japan, and migrates to the Philippines and Malaysia for the winter. With its bell-like call, its return to the northerly latitudes is hailed as a harbinger of spring.

△ *Protest. A young flamed minivet launches a loud squawk of protest from his tree stump. These birds are rarely silent but their noise is usually confined to their incessant chattering. Overleaf: A pair of minivets, **Pericrocotus wrayi**, make a colourful portrait. They are usually seen as just flashes of red and yellow flitting through the forests.*

Insect and spider food

Minivets' food is insects and spiders picked off the foliage of trees. Insects are sometimes caught on the wing in the manner of a flycatcher, or the bird may hover over a flower to take its victims.

Cup nests

Minivets build a cup-shaped nest of twigs, rootlets, moss, grass or pine needles, bound together with spiders' webs and sometimes decorated with lichens. The nest is usually anchored near the top of a forking branch, 15–20 ft from the ground. In it are laid 3–5 eggs, pinkish, greenish-white or bluish-grey marked with brown, grey or purple spots. The female alone incubates the eggs but both parents feed the chicks.

Any silk will do

It is sometimes suggested that a major difference between man and animals is that we can exploit our environment. It is, however, only a question of degree. Some birds, to give a simple example, show great ingenuity in collecting materials and fashioning them into a nest. One instance which provokes admiration is the way they will use the silk of spiders' webs and, using only their beak, weave it into the nest to give maximum strength with flexibility. Small birds like minivets tend to use this material. Hummingbirds use it a great deal, and the nests of small hummingbirds may sometimes be made of nothing but spiders' silk. The adaptability of birds was illustrated a few years ago more especially when tangled plastic strands as fine as spiders' silk was sold to soft fruit growers to put on their currant bushes. It was intended to protect the buds from attacks by bullfinches, notorious for stripping fruit buds in spring. The idea was that it would form a thin net over the bush which would keep the birds away. In at least one place in England, soon after this plastic silk was put on the market, bullfinches were seen flying away with it in their beaks to weave into their nests. One pair of bullfinches was seen making regular trips from the currant bushes in a market garden to their nest 100 yd away, to make the fullest use of this heaven-sent bounty. The nests of minivets are so small they look like a knot on the branch and the lichens bound on by spiders' web camouflage it perfectly, so they are almost impossible to find except by watching the movements of the minivets.

class	**Aves**
order	**Passeriformes**
family	**Campephagidae**
genus & species	***Pericrocotus flammeus*** *flamed minivet*
	P. igneus *fiery minivet*
	P. roseus *rosy minivet*
	P. solaris *mountain minivet*
	P. divaricatus

1615

Mink

There are two species of these valuable fur-bearers: the American mink and the European mink. They are close relatives of the stoat and weasel and belong to the family Mustelidae which includes the badger and marten. The mink is very similar in appearance to the stoat. Males are 17—26 in. from the snout to the tip of the tail, which is 5—9 in. long. Females are about half the size of the males. The ears are short and set close together, the tail is bushy, and the toes are partly webbed. Wild mink are light to dark brown with a white patch on the lower lip and chin and a few white spots on the belly, but a number of colour varieties, from white to almost black, have been bred by artificial selection on mink farms.

The American mink is found in most parts of North America, from the Arctic Circle southwards to Mexico. It is absent from the southern half of Florida and parts of the southwest United States. The European mink is much rarer than it was formerly and it is difficult to tell exactly what is its present range because the introduced American mink is so similar. It is still found in Russia, parts of eastern Europe, Finland, possibly France, and within the last century it has spread across the Urals into the Siberian plains.

Riverbank dweller

Mink hunt on land and, like weasels, kill more than they need, but they also hunt in the water, like otters. Mink live along the banks of wooded streams, rivers, marshes and lakes. Being nocturnal, they are rarely seen, but their droppings, footprints and remains of prey that are found along the bank are sure evidence of mink.

Male mink wander long distances while females have small home ranges. Each mink has a den among rocks, in holes in trees or in burrows of other animals. In North America they take over woodchuck or muskrat holes and in Britain they enlarge the burrows of water voles.

▽ The silent watcher. Unlike their relatives, mink are equally at home in water and on land.

Miserly mink

Mink prey on both land and water animals. In water, they feed on crayfish and frogs as well as fish, including salmon and trout. On land they catch many kinds of small mammals including muskrats and water voles, and birds such as moorhens, ducks and other birds that nest on the ground. They also raid poultry runs. Surplus food is stored in the den; one in the United States contained 13 muskrats, 2 mallards and a coot.

Wandering lovers

Mating, which takes place from February to March, is the only time when minks make much noise; both sexes purr. Each male may mate with several females as he passes through their territories but he usually stays with the last one. As with all members of the mustelid family, there is delayed implantation of the embryos and the kits are not born until 45—50 days later, although it may be any time between 39 and 76 days. There are usually 5 or 6 kits in a litter.

The kits stay in the nest of fur and feathers for 6—8 weeks, although they are weaned at 5 weeks. They are very playful, sliding down banks like otters and indulging in noisy, rough-and-tumble mock fights. Until autumn they accompany their parents on hunting trips, then the family breaks up and each leads a solitary life until the next mating season.

Introduced vermin

Mink is one of the most valuable fur bearers and has been bred commercially since 1866 either on large farms or ranches or in backyard cages as a source of extra income. In 1951 the pelts of two million ranch-reared mink were sold on the United States market. Farms have been set up elsewhere from Iceland to the Falkland Islands, and in Russia American mink have been introduced to replace the vanishing European mink, which, anyway, has an inferior pelt.

In many places mink have escaped from farms and colonised the surrounding country. In Iceland they have attacked nesting ducks and waders, with severe effects. In Scandinavia they have turned their attentions to salmon. The first mink farm in the British Isles was established in 1929 and soon feral mink were being recorded. Sometimes there were reports of pine martens in southern England, but whenever it was possible to follow up reports they were found to be mink. Wild mink were first recorded in one area when a fur dealer was sent a mink pelt in mistake for an otter pelt.

It was originally thought that mink would never become established in Britain but in

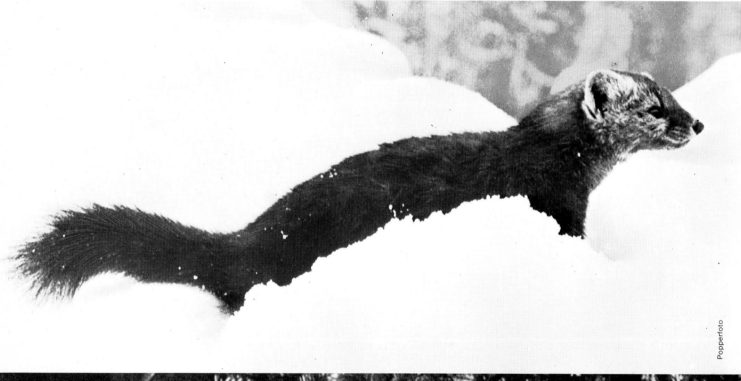

Popperfoto

Popperfoto

1956 they were found breeding in Devon. Signs of breeding were later found in other parts of England as well as Wales and Scotland and mink have spread rapidly along many river systems. They have not become a pest as was once feared, neither have they affected significantly the populations of other species.

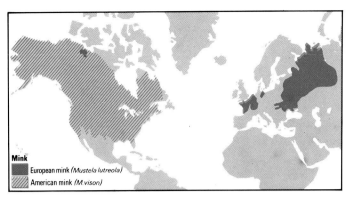

Mink
European mink (*Mustela lutreola*)
American mink (*M. vison*)

class	**Mammalia**
order	**Carnivora**
family	**Mustelidae**
genus & species	***Mustela lutreola*** European mink *M. vison* American mink

◁ *Cutting a figure. Profile of a mink as it slinks through the snow.*
◁▽ *Platinum mink, a variety of American mink from selective breeding, is bred on farms for the colour of its fur. It does not occur in the wild.*
▽ *Inquisitive bystander. Eyes glued and all senses alert, an American mink pulls itself up onto its front legs and cranes its neck for a better view of the world around.*

Jane Burton: Photo Res

Minnow

Although this name in its strict sense is used for a small European freshwater fish it has been widely used for many large-headed fishes. These cannot be ignored, so here we shall include with the original minnow a number of related fishes of the carp family.

The minnow is one of the commonest and best known freshwater fishes in Europe and it extends into Asia from Siberia to Lake Baikal. It has a cylindrical body, seldom more than 4 in. long, marked with dark bars and covered with small scales. There is a single, small, dorsal fin and the anal and paired fins are small and slightly tinged with red. The colour is brownish-green to silver grey, to silvery white on the belly. Minnows live in the deeper parts of clear brooks and streams where there is a sandy or gravelly bottom. Spawning is in May to July when the male, slightly smaller than the female, becomes almost black, with a scarlet mouth, belly and fins, and has whitish tubercles on the head. These growths are sometimes called pearl organs.

Many minnows

There are many kinds of minnows, some are active swimmers, some lethargic, some are small, some large. They are found in both temperate and tropical waters, where they occupy many kinds of habitat, from small brooks and ponds to broad rivers, mountain streams, glacial lakes and warm springs. They all agree in certain features: they have no teeth in their jaws but have strong throat teeth; their fins have soft rays; the pelvic fins are set far back on the body, and in the breeding season the males in many species, sometimes the females also, develop tubercles on the head.

Many species live in North America where they are known under a variety of names: squawfish, fathead, split tail, hardhead, shiner, fallfish, stoneroller. Most of them are used as bait and some of the larger species are used as food fishes. In parts of the United States there are minnow farms, where small fishes are cultivated to be sold for livebait. One that is used for bait is the common shiner, up to 8 in. long but very like the European minnow in shape. The split tail, with a deeply forked tail, of California, grows to 12 in. and the hardhead of the same area may be up to 3 ft.

Chub and dace

The chub's name refers, in several European languages, to the shape of its head, and 'chub' almost certainly refers to the chubby cheeks. The chub of Europe and southwest Asia is a surface-living fish up to 2 ft or more long and commonly weighing up to 7 lb, but up to 12 lb has been recorded in Continental rivers. In North America there are over a score of chubs, most of them less than 4 in. long, with the flathead chub going up to 12 in. The dace, of Europe and parts of Asia, closely resembles the chub but with a

DB Lewis

more slender body and a more forked tail fin. It is up to a foot long and very rarely more than 1 lb weight. It lives in clear streams with rapid water and, like the chub, is often found in trout streams. The bleak, up to 6 in. long, can be mistaken for a small dace. Shoals of bleak often feed at or near the surface in slow rivers. There are several dace in North America and some of these are brightly coloured, like the southern red-belly dace, 3 in. long, with a red stripe with dark borders on each flank.

The doctor fish

Another well-known minnow-type is the tench of Europe and western Asia, distinguished by its two barbels and golden-yellow scales. It lives among water plants in quiet ponds with muddy bottoms. Normally slow-moving, it passes the winter in a torpid state in the mud. Up to 28 in. long and 8 lb, rarely 17 lb, weight, it is considered a tasty fish and has been introduced into the United States. The tench has been called the doctor fish. This name came from the belief that injured fishes touching it will have their wounds healed, by the curative properties of its slime.

One American minnow, the stoneroller, is remarkable for its very long intestine which is looped several times round its swimbladder.

Mound builders

Most minnows and their near relatives feed on small prey such as water fleas, freshwater shrimps and insect larvae, but others feed on smaller fishes. Their breeding habits are even more varied. Many, including the common European minnow, shed their eggs onto gravel or sand or among water plants and show no parental care. In a number of species the males dig shallow pits in the sand for the females to lay in, or the females lay their eggs on the under surfaces of stones or submerged logs. A few species gather small stones and pile them to make nests for the eggs and the most remarkable of these is the fallfish. The males, smaller than the females, and never more than 18 in. long, move stones as much as 3 in. diameter, making nests 6 ft across and 3 ft high.

Minnows are fair game

The enemies vary with the size of the fishes and the situations in which they are living. The fact that so many minnows are used as livebait tells its own story. In the wild these small fishes are the prey of waterside birds such as heron, bitterns or kingfishers as well as of larger fishes. The tench, for example, is soon cleaned out of ponds by other fishes, especially pike, unless there is a good growth of plants in which they can

Popperfoto

Heather Angel

Top: Common minnow. This small fish is used by many anglers as bait. A trick of some anglers is to paint the inside of their minnow-can white, the minnow assumes the lighter colour and so is more conspicuous to pike and perch in deep dark water.
*Centre: Dace, **Leuciscus leuciscus**, a silvery coloured fish. It frequently feeds from the surface of the water.*
Bottom: Tench. A large fish that lives in quiet ponds where the bottom is muddy.

hide. The common minnow is particularly vulnerable when it gathers in shoals over shallow sand or gravel banks, each female laying 1 000 small sticky eggs. On the continent of Europe these shoals are the occasion for a regular fishery, the minnow being used for food. They are good to eat if caught in sufficient numbers, and they used to be served at table in England. In 1394, for example, 7 gallons of minnows were served at a banquet given by William of Wykeham for King Richard II.

Danger signals

Because minnows are numerous, easily caught, and easy to keep in aquaria, it is natural they should be used as laboratory animals. One thing about them that has been especially studied is their reaction to danger. When one of the shoal is seized by a predator the rest of the shoal bunches together, then swims away, and does not come back to the place for a long time. In the laboratory they do the same if an injured minnow is put among them. Further study showed that a substance given out into the water when the skin is broken or cut is smelt by the rest, causing them to behave this way. They will do the same if a little juice is extracted from a piece of minnow skin and, in dilute solution, is poured into the water. Tests showed that extracts from the intestine and liver had no effect, that extract from muscle had only $\frac{1}{20}$ the effect of skin extract and from the gills $\frac{1}{5} - \frac{1}{10}$ that of skin extract. Further tests revealed that minnows showed this fear reaction to extracts from the skin of other species of minnow, but to a lesser extent than to that from skin of their own species, and that extracts from the skin of fishes in other families had little or no effect. Moreover, if a minnow dies of injuries and its body lies in the water, this 'alarm-substance' continues to be given out for several days.

class	**Osteichthyes**
order	**Cypriniformes**
family	**Cyprinidae**
genera & species	***Campostoma anomalum*** *stoneroller* ***Chrosomus erythrogaster*** *red-belly dace* ***Hybopsis gracilis*** *flathead chub* ***Leuciscus cephalus*** *chub* ***L. leuciscus*** *dace* ***Notropis cornutus*** *common shiner* ***Phoxinus phoxinus*** *common minnow* ***Semotilus corporalis*** *fallfish* ***Tinca tinca*** *tench* *others*

▽ *Lively, colourful, good community fish, the white cloud mountain minnow* **Tanichthys albonubes** *is popular with aquarists.*

Jane Burton: Photo Res

Mite

Mites are minute relatives of spiders and scorpions. They are the small members of the order Acarina, the largest being the blood-sucking ticks which will be described later. The body is rounded, usually about $\frac{1}{2}$ mm long, with no visible division into two parts as there is in the bodies of spiders. As in spiders, there are four pairs of many-jointed legs with claws at the tips, and a pair of chelicerae. These may be sharp and able to pierce the bodies of other animals. In spiders the chelicerae are the poison fangs. On each side of the chelicerae are two short limbs called pedipalps, which are not pincers as in scorpions or false scorpions (p. 886), but are leg-like and carry many sensory hairs.

The total number of mites is not known but there may be in excess of a million species. There are only a handful of scientists studying mites compared with the thousands studying insects, yet in 1966 descriptions of 1 000 new species and 150 new genera of mites were published. Little is known of the habits of many mites and new mites are being discovered all the time. Some, however, have as great an economic importance as insects and are now being closely studied.

Mites almost everywhere

Mites are found in all sorts of odd places and have a great diversity of habits but they are readily overlooked because of their small size. They are found in the nostrils of seals, among the gills of crayfish and the hearing organs of moths. Others cause plant galls, a few live in the sea and many eat decaying matter, such as the cheese mite that feeds on decaying cheese. Mites are also found in the Antarctic where they feed on fungi and decaying plants in the scanty patches of moss and algae. Perhaps the best known mite is the harvest 'bug' or harvest mite. The larval stage of the harvest mite waits in matted vegetation or low bushes and clings to warm-blooded animals as they pass. The larvae of harvest mites have only six legs: they are white to orange-red and about $\frac{1}{100}$ in. long – just visible to the naked eye. They pierce the skin of the host with the chelicerae, which is painless, but a fluid is injected to break down the cells of the skin to allow liquid food to be sucked up. This causes quite severe irritation and small inflamed pimples. As the harvest mite sucks up food its body stretches until it is three or four times its original size. The larval mite then drops to the ground and completes its life cycle, changing first to an eight-legged nymph, then to the mature adult.

The harvest mite is picked up by people without adequate protection from clothing when they walk or sit in vegetation where these mites are abundant. Another mite that attacks humans is the itch mite which burrows in the skin. The eggs are laid in the burrows and larvae emerge 2–3 days later. Scratching the sources of irritation increases the risk of infection. Some mites

*Mites are found just about everywhere. The brightly coloured mite (below) is a water mite **Hydryphantes** that lives in fresh water. The adult, shown, is predatory and feeds on crustacea, insect larvae and any other small aquatic animals. The larvae are parasitic on aquatic animals. On the legs of this mite are long hairs which help the animal in its rapid swimming movements. (30 × life size)*

J AL Cooke

transmit disease directly, such as scrub typhus, and some very minute mites that live on dandruff and sloughed skin trigger asthmatic attacks when they are inhaled with household dust.

Not always dangerous

Infestation by mites often causes little or no harm to the host except the loss of some body fluid. Some mites infest the 'ears' of certain moths, such as the army worm moth a severe pest in North America. They rupture the eardrum of the moth, lay their eggs on the ear duct and pierce its wall to suck the moth's blood. No doubt the 'ear' is put out of action, but in a study of army worm moths it was found that only two out of every 1 000 had mites infesting both ears. No-one knows why all the mites should head for only one ear but it does prevent the moths becoming deaf.

Gradual growing up

The mating behaviour of mites is very similar to that of spiders, in that the male transfers a bag of sperm, the spermatophore, to the female. In one species studied in the laboratory, the male climbs onto the female, feeling with his legs and pedipalps, then turns her onto her back and exudes a spermatophore. This is transferred to her genital opening and gradually drawn in.

Mites lay large eggs one at a time at long intervals. The mite that lives in the 'ears' of moths lays eggs almost half its own size at intervals of 2 hours or more until about 90 have been laid. Before laying the female mite scrapes a patch on the lining of the 'ear' on which she sticks the egg. Each time an egg is laid the process is repeated until there are several neat rows of eggs.

The larvae that emerge from the eggs have only three pairs of legs or, in the case of the gall-forming species, two pairs. At intervals the larvae shed their skins and each time they emerge they look more like the mature adults.

Pesticides cause pests

One of the drawbacks to the powerful pesticides developed in the last 20 years or so is that they sometimes 'backfire' by making new pests. This does not mean, of course, that new animals are created. A pest is an animal that causes sufficient damage to man's property or person to be of economic importance. Such a pest is the red spider mite, which attacks fruit trees and other crops. The red spider mite weaves sheet webs on plants that look like opaque cellophane. The mites live in large numbers within the webs, sucking the plants' juices. At one time it was not a pest because the damage it did was negligible. Then the widespread application of powerful pesticides killed off the red spider mites' enemies, resulting in an astronomical increase in these mites.

As a result pesticides had to be employed against the red spider mites, but unfortunately they became immune and even larger amounts had to be used. Such enemies as were left, if they did not succumb immediately to the poisons, died from eating poisoned mites.

There is now hope that the red spider mite may be rendered innocuous again. The answer seems to be in biological rather than chemical control. The red spider mite is a very serious pest of cucumber and other greenhouse crops, but experiments have shown that their numbers can be drastically reduced by introducing a predatory mite *Phytoseiulus riegeli*. Paradoxically it is sometimes necessary to infect a greenhouse with red spider mites to ensure an adequate supply of food for the predators. The system works well, it is apparently 10 times cheaper than chemical control, and there is the advantage that an animal cannot become immune to another that eats it.

phylum	**Arthropoda**
class	**Arachnida**
subclass	**Acari**
genera & species	***Tetranychus urticae*** *red spider mite* ***Trombicula autumnalis*** *harvest mite* ***Tyrophagas putrescentiae*** *cheese mite, others*

small and spidery . . .
. . . big and beautiful

▷ *Not quite the colour you might expect: adults of a red spider mite* **Metatetranychus ulmi***, eggs and hatched egg cases on the underside of an apple leaf. It is a pest of gardens and orchards; infestation is easily recognizable from the loose webs covering leaves on which the eggs are laid (30 × life size).*
▽ *The larvae of the velvet mite* **Allothrombium** *are parasitic on this freshly moulted scorpion. The larvae feed on the scorpion's blood and are not above using the unfortunate host as a way of getting about.*

Jane Burton: Photo Res

◁ *A common sight.* **Parasitus coleopterorum** *nymphs clinging to the underside of a dor beetle which is struggling to turn itself the right way up. These mite nymphs, each one about 1 mm long, are only using the beetle as a means of transport, and do not do any harm to the beetle. When the nymphs want to moult they will drop off from the beetle to do so.*

▽ *The giant red velvet mites* **Dinothrombium** *are, as their name suggests, large, about 1 cm long. They are in fact the largest of the Acarina. Their bodies have the exact texture of velvet. They are mainly found in desert and semi-desert areas and also in some humid parts of the tropics. After heavy rains they emerge in such great numbers that they cover the ground and can be counted in their thousands.*

Anthony Bannister: NHPA

Mockingbird

The mockingbird's ability to mimic the songs of other birds has made them extremely popular in the southern United States. They are the 'state bird' of five states—Arkansas, Florida, Mississippi, Tennessee and Texas. Most mockingbirds are rather dull in appearance. They look like thrushes, but belong to the same family as catbirds (p. 525). They are 10–12 in. long with more slender bodies than thrushes. Their tails are longer and their bills more slender. The best known mockingbird is the northern mockingbird of the southern United States. It has flourished with the advance of civilisation and has spread northwards to New England. The northern mockingbird is grey, lighter underneath and almost slate-grey on the wings and back. The throat is white and there are streaks of white on the wings. Other mockingbirds live in the West Indies, and in Central and South America as far south as southern Argentine and Chile. The two blue mockingbirds of Mexico are unusually brightly coloured. One is all blue with a black eye-stripe while the other has a white belly. In the Galapagos Islands there is a mockingbird that is rather different from the others, having a larger, more compressed bill and longer and stronger legs. The Galapagos mockingbird has evolved in isolation in the same way as Darwin's finches (p. 751) and there are now several forms.

Vigorous defenders

The ever-changing song of the mockingbird is heard all the year round. This is because they use it to maintain territories throughout the year although outside the breeding season both males and females occupy separate territories. There is a record of a pair of mockingbirds that bred together for several years in succession, but after raising each brood they defended their individual territories against each other. The next spring hostilities ceased and the pair reunited. Fights between rival mockingbirds are rare and boundary disputes are settled by a 'dance'. The two birds face each other with their heads and tails raised and dart backwards and forwards across the disputed boundary.

Mockingbirds have attracted attention not only because of their prowess at mimicry, but also because of the vigour with which they defend their territories. They swoop at dogs, landing on their backs and pecking them, and attack men, cats, squirrels, snakes and other possible enemies. They also have a threatening display in which they face their enemy, cock their tail and utter a loud, sharp note, while fanning the tail. Young mockingbirds perform this display but at first they do it indiscriminately, not directing it at any particular object. They have to learn from their parents what animals are potentially dangerous, and they display at these.

Galapagos mockingbird in a prickly situation.

Fruit and egg thieves

Mockingbirds are not so popular in their choice of food. They are sometimes persecuted for eating fruit such as grapes and oranges, but it is unlikely that they ever cause much serious damage. They eat a lot of wild fruit such as holly, blackberry, poison ivy and elderberry and many insects, including harmful ones such as cinch bugs, cutworms and boll weevils. Large numbers of beetles and grasshoppers are also devoured. The Galapagos mockingbird has formed the habit of cracking the eggs of swallow-tailed gulls and sipping the contents.

Rapid upbringing

A mockingbird's nest is constructed with feverish activity and is sometimes completed in as little as two days, but more often it takes 3–4 days. The nesting begins in April or May in the United States. Both sexes work together to make a strong, cup-shaped nest of small twigs, lined with grass and rootlets. It is usually well hidden in a bush or tree 3–10 ft above the ground, rarely higher. The mockingbirds often nest near houses, in garden shrubs or in vines climbing the side of a house. They lay 3–6 prettily coloured eggs which are bluish white to rich blue, or green, with brown spots and blotches. The female mockingbird does most of the incubation and the male has only been seen to incubate for a few minutes when the female has left the nest. On her return she may even drive him off the nest.

The chicks hatch out after 12–14 days and leave the nest about 2 weeks later, so the whole process from nest-building to the chicks' flying takes little over one month and two or three broods may be raised in one year. Both parents feed the rapidly growing young, mainly on insects. If one adult arrives while the other is at the nest it waits its turn until the other has gone, then comes forward to feed the nestlings.

Many-tongued mimic

Although Linnaeus, who compiled the modern scientific method of naming animals and plants, had never seen a mockingbird, he called it *Mimus polyglottos*, the 'many-tongued mimic', for it is this characteristic that has made the mockingbird famous. Its song is a burbling, wren-like melody in which the pattern of notes and syllables is continually changing. It usually consists of a series of notes, each repeated several times. Perhaps 10% of these, at the most, will be mimicked sounds, which change from time to time; mockingbirds sing certain favourite tunes very frequently, then discard these for new ones. In addition to the natural song, mockingbirds imitate other bird-calls, human voices and mechanical sounds. Their mimicking has, however, been exaggerated and reports of a single mockingbird imitating 20 or 30 other species are probably overstatements. It has been suggested that one cause of this exaggeration is that the mockingbird's natural repertoire is so varied and changing that it probably makes sounds resembling those of other species quite by chance.

There has been much speculation why birds mimic other birds. Chaffinches (p 542) learn the songs of their parents and neighbouring males by imitation and it may be that this method of learning is very highly developed in mockingbirds, parrots and others, so they continue to pick up other songs and noises throughout their lives. The imitations are certainly very precise. When nightingales were brought from Europe and kept at the Singing Tower in Florida, the local mockingbirds picked up their song. An analysis on a sound spectrograph showed that the mimicked song was an exact imitation of the nightingales' song.

*Fruit thief caught in the act, **M. polyglottos**.*

class	**Aves**
order	**Passeriformes**
family	**Mimidae**
genera & species	*Melanotis caerulescens* blue mockingbird **Mimus polyglottos** northern mockingbird **Nesomimus trifasciatus** Galapagos mockingbird

Mole

There are several species of true moles, all in the northern hemisphere. Of these, the European mole has been the most completely studied. Its body is cylindrical, 5¼ in. long, the female is slightly smaller than the male, with a club-shaped tail 1¼ in. long. The snout is long and pointed, the eyes are very small and hidden in the fur and the external ear is no more than a ridge. The head and snout are beset with long bristles. The velvet fur, dark grey to almost black, is short, dense and without set. Occasionally it is ash-grey, yellow, orange, cream or white. All four limbs are short and enclosed within the skin of the body. The forelimbs are well forward. The front paws are broad, with 5 toes and an extra crescentic bone giving even greater breadth. Each toe has a long, strong claw. The hindfeet are small by comparison, but not weak as they are usually described.

The common or European mole ranges across Europe and much of Asia. In southern Europe and southwest Asia it is replaced by the Mediterranean mole and in southeast and eastern Asia, including Japan, it is replaced by the eastern mole. In America there are a dozen moles and shrew moles, belonging to a separate sub-family, and another, the star-nosed mole, is in a third subfamily.

Life below ground

Moles are restless creatures; busily, and almost excitedly, throughout the day and night, they hunt and feed for 4½ hours then rest for 3½ hours. The natural habitat of the European mole seems to have been woodlands but as these areas shrank with more land being tilled, moles invaded fields. They live wholly underground, seldom coming to the surface and then for only short spells. Everything in their bodies is adapted to a burrowing life. The eyes are of little use and the chief senses are those of smell and hearing, and they almost certainly have an ability to detect vibrations which is a kind of 'touch at a distance'. The forepaws which are directed face outwards, are permanently extended and can at most be only partially closed. They are used alternately in loosening the earth being brought forward so the tips of the claws are beyond the tip of the snout and then pulled backwards, to loosen earth and throw it behind. The relatively weak hindfeet are used for pushing the body forward, but the mole moves through its tunnels in a half-swimming, half-looping gait.

Surface runs are often used, the soil being heaved up into a ridge as the mole drives its way along with its back about ¼–½ in. below the surface. There is, however, a system of permanent tunnels. At 3–6 in. below the surface is a horizontal network, with another 12 in. below that, the two being connected by vertical or oblique shafts. From the lower network there are occasional shafts running down in an irregular zig-zag to end blindly 3–4 ft below the surface.

Coming out of his underground burrow the mole stretches his legs and takes in the fresh spring air.

Joe B Blossom: NHPA

This plan will vary slightly according to the nature of the subsoil. Earth, dug out in extending old tunnels or making new ones, is thrown out at the surface as molehills, which are up to 1 ft across and 6 in. high. Moles were thought to throw this earth out with their snouts but it is now known that they push it out with their forepaws.

Stores of earthworms

A mole cannot survive more than a few hours without feeding and when earthworms, its chief food, are plentiful it may store them. It bites off the tip of the worm's head end, twists it into a knot and pushes it into a cavity in the soil. These stores sometimes include hundreds, even thousands of earthworms. Should the mole not need them the worms in time re-grow their head ends and burrow away. When eating a worm the mole holds it down with its forepaws, bracing the body with the hindfeet set well apart, and chews it from the front end backwards. It also eats earthworm cocoons, as well as insects, especially leather-jackets, wireworms and cutworms, all pests of agriculture. A single mole will eat 40–80 lb of food in a year, foraging over ⅒ of an acre, on average, but there is a horizontal migration to new feeding grounds in autumn. It does not need to drink when feeding entirely on earthworms since they contain 85% water.

Underground nurseries

A mole often enlarges part of a tunnel to make an oval nesting cavity of either grass, leaves or twigs, about 1 ft long. There is often a complicated system of tunnels around this which gives the occupying mole several bolt holes. This complicated network has led to the romantic idea that a mole rests within its 'fortress'. The cavity is usually at the surface under a mound of excavated soil, which may be up to 8 ft in diameter and 3 ft high, but many nests are below the surface to a depth of up to 3 ft, with no soil on the surface to give away the position. Males and females make fairly similar nests for resting, but a female's is said to be smaller. When she is soon to give

birth, she makes a separate nursery nest, sometimes making a fresh one for each new litter. The boar and sow are together only when mating, at the end of March and beginning of April. The young, born 5–6 weeks later, are blind, naked and pink at birth. There are from 2 to 7 in a litter but 3 or 4 is the usual number. The skin soon turns a bluish-slate and the fur comes through at 2 weeks. The eyes open at 22 days, by which time their weight has increased from ⅒ oz at birth to 2 oz. The young leave the nest at 5 weeks and it is then that they spend more time than usual at the surface. They are sexually mature the next February. The lifespan is 3 years.

Dangerous infancy

Enemies include tawny and barn owls, herons, weasels, stoats, badgers, foxes and cats. Common rats kill moles in their nests. Moles are especially vulnerable to tawny owls when the young come above ground. At times they may form 50% of the tawny owl's food. In eastern Europe moles are a large part of pine martens' food.

Delicate touches

Recent researches have shown how large a part is played in a mole's life by its organs of touch. Some of these may be so delicate they serve as 'touch at a distance', that is, detecting minute vibrations in the soil which may come from other animals, even as small as worms, moving some distance away. Little is known about this sensitivity but a mole's skin has more touch organs than that of any other mammal. There are thousands of tiny papillae, known as Eimer's organs, on the tip of the snout. Each papilla has a sensitive hair embedded in it. There are sensitive hairs on the tail used as organs of touch when a mole is moving backwards through its tunnels, which it can do easily as the fur has no set. (Whichever way you stroke it the fur does not look untidy.) A mole also has sensitive patches of skin on its body, especially on the belly, known as Pinker's plates.

The most extraordinary organ of touch is

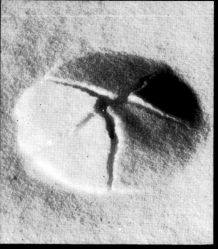

◁ 'So he scraped and scratched and scrabbled and scrooged, and then he scrooged again and scrabbled and scratched and scraped, working busily with his little paws and muttering to himself, "Up we go! Up we go!" till at last, pop! his snout came out into the sunlight.' This excerpt about the Mole taken from 'Wind in the Willows' is a delightful description of the activities of any mole coming up from his underground home. Although the mole is not actually visible (top left) the cracks in the snow and the final break through are definite indications of life below. A mole has surfaced (centre left) but it will not be long before he is tunnelling underground again.

▷ These pictures earned Peter Stafford the title of Wildlife Cameraman of 1967, for they are the first photographs ever taken of a mole underground. In 'Wildlife and the Countryside', February 1968, the photographer explains how he had to wait patiently to get a picture of the mother with her babies. His patience was rewarded as we see. He found the four babies, which were only about 2 weeks old (above), in a nest of dry grass 3 feet below a Sussex field. He had difficulty in getting a picture of the mother as she never showed herself. When she realised her nest was disturbed she would quickly cover her young by throwing up soil over them from below. After a few days she changed her technique and threw soil from the side. This was the photographer's chance; he cleared the side tunnels of loose earth and waited. Then suddenly she was there, her head and shoulders, with fur brushed backwards, appeared from the side of her tunnel (below). The sound of the camera shutter made her disappear again very rapidly and she covered her babies once more, but the photographer had his winning picture.

◁ Gateway to Clumber Park, a beautiful building in Nottinghamshire that now belongs to the National Trust. The house, most of which has been pulled down, was originally owned by the Duke of Newcastle. It is a very popular place for visitors and, so it seems, for moles. As the woodlands, the natural habitat of the European mole, are slowly decreasing in size the mole is excavating his tunnels in fields, gardens and grassland, consequently making many unsightly molehills wherever he goes.

in the star-nosed mole of North America. This is similar to the European mole in shape and habits except that it has a tail nearly as long as its body and that it gets most of its food in water. It swims well and catches insect larvae, freshwater shrimps and small fishes, but it also eats earthworms. Around the tip of its muzzle is a ring of 22 pink, fleshy rays or 'tentacles', forming a star. When searching for food these are in constant movement except for two upper tentacles held rigidly out in front.

class	**Mammalia**
order	**Insectivora**
family	**Talpidae**
genera & species	***Condylura cristata*** *star-nosed mole* ***Talpa caeca*** *Mediterranean mole* ***T. europaea*** *European mole* ***T. micrura*** *eastern mole, others*

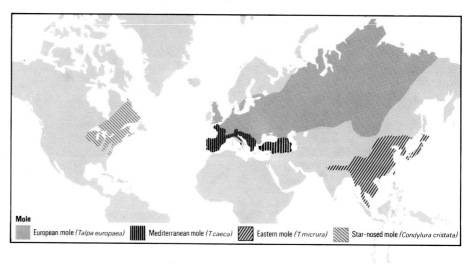

Mole

	European mole *(Talpa europaea)*		Mediterranean mole *(T.caeca)*		Eastern mole *(T.micrura)*		Star-nosed mole *(Condylura cristata)*

Most unusual sight—the star-nosed mole above ground. This mole lives almost entirely underground, searching constantly for food with the sensitive feelers at the end of its nose.

Lynwood M Chase

△ *Palestine mole-rat, the most mole-like of the mole-rats, has invisible eyes and tail.*

Eric Hosking

Mole-rat

There is an obvious link between the subterranean habits and molelike appearance of mole-rats. Most have short, dense fur, small external ears and short, but strong feet. All these features allow them to run down their burrows easily.

Most mole-rats live in Africa but the 6—12in. long Palestine mole-rat of the family Spalacidae lives in the eastern Mediterranean from Libya to Greece, and in the plains of southern Russia. It is the most molelike of the mole-rats. The tail and eyes are invisible and the external ears are merely small flaps. The dense, grey fur can be brushed in almost any direction. Most of the mole-rats of the family Rhizomyidae live in eastern Africa from Ethiopia to Tanzania, but some live in southeastern Asia where they are called bamboo rats. They are rather volelike, resembling the American pocket gophers. The third family which includes the blesmol, is the Bathyergidae, found in many parts of Africa south of the Sahara. They are the most ratlike of the mole-rats, and their incisors, which are long in all mole-rats, are very prominent. These mole-rats, 4—7½ in. long, with a tail up to 1 in. in length, may be any shade of brown, from a light yellowish colour to almost black. One species, the naked mole-rat, appears completely naked, but there are quite a few long fine hairs scattered over the body.

Extensive burrows

Mole-rats dig galleries of burrows which may be so extensive as to make it impossible to ride horses over mole-rat infested ground, the ground collapsing underfoot. In some places they have even undermined railway sleepers, so they have sunk as a train passed over. The burrows are usually about 6 in. under the surface but in sandy soil they may go deeper. The position of the burrows is indicated by mounds of earth, often very much like those of true moles, but in dry weather, when the ground be-

comes hard, the African blesmol redistributes earth around its tunnels rather than force it up in mounds. Some mole-rats live in colonies of a dozen or so individuals which share a nest and burrowing activities. There are usually food stores leading off tunnels near the nest, and the Palestine mole-rat also has sanitary chambers near the nest which are walled-up when full. This mole-rat lives in places liable to flooding, where it builds large nesting mounds, sometimes 9 ft. in diameter and over 3 ft high, so they can escape from waterlogged ground. In the dry season, when they are not breeding, they build smaller mounds, where they sleep. The blesmol also makes its nest and storage chambers in relatively high ground, and intricate networks of tunnels are often to be found in termite mounds.

▽ *As naked as nature intended! The naked mole-rat has but a few fine hairs on its body.*

▽▽ *Naked mole-rat's hole. Excavation resembles a miniature volcanic eruption.*

Des Bartlett: Photo Res

Des Bartlett: Photo Res

Underground pest

Mole-rats are vegetarian and as such are often great pests in fields and gardens, although in some places they are useful in that their burrows help the ground to absorb moisture during the rains which would otherwise run off the surface. Mole-rats eat a variety of plant material, mainly roots, bulbs, rhizomes and other underground parts. The blesmol eats large quantities of the fleshy bulblike bases of grasses and the Palestine mole-rat sometimes comes onto the surface at night to eat grasses and seeds. Both species store food in chambers near the nest for use when the ground is flooded or too hard to dig. Some insects and other small animals are also eaten.

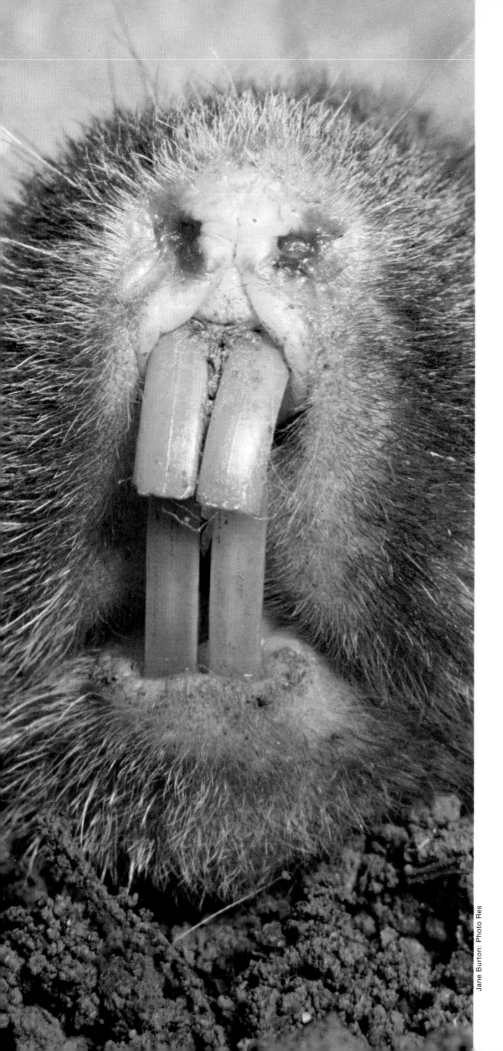

Biting babies

Mole-rats breed in the wet season, when food is abundant. The Palestine mole-rat builds its large breeding mound between October and January. After mating the male leaves the female and the 2–4 babies are born a month later. At first they are pink, naked and helpless, unlike the new-born blesmols which can inflict painful bites. At 4–6 weeks, they leave the nest. This is the time when Palestine mole-rats are most often found above ground, for young mole-rats often travel overland at night in search of a vacant piece of ground.

No safety underground

Even in their burrows mole-rats are not safe from enemies. They are dug out by honey badgers, foxes and jackals, and even eagle owls have been seen to catch mole-rats by watching for the disturbance caused by their working, pouncing and driving their talons down through the earth to reach them. Owls hunt mole-rats when they come to the surface at night and there is a particularly heavy toll taken when flooding forces mole-rats out of the ground.

Ways of digging

Although closely resembling true moles, mole-rats belong to a different order of mammals. They are rodents, or gnawing animals, and several species of mole-rat actually gnaw their way through the earth, whereas the true moles thrust with their snouts and scrape with their spade-like forefeet. The blesmol gnaws at the soil and pushes the loose earth under its body and out behind. When a certain amount has accumulated it forces the earth back along the tunnel towards an opening. The loose earth is compacted and extruded above the surface as a pillar about 6 in. high and 2 in. in diameter. Its relative, the naked mole-rat, kicks earth out of the tunnel, looking as if a miniature volcano is erupting. After it has loosened the earth, it scrapes it back and throws it out by supporting its body on forefeet and tail and kicking violently with its hindfeet. The Palestine mole-rat also digs with its teeth. It has very powerful jaw muscles which almost fill the orbits, the cavities in the skull where the eyes lie. In most animals the eyes fill the orbits but in this mole-rat they have virtually disappeared. The tough snout is also used to compact loose earth against the sides of the tunnel.

class	**Mammalia**
order	**Rodentia**
family	**Spalacidae**
genus & species	***Spalax ehrenbergi*** *Palestine mole-rat*
family	**Rhizomyidae**
genus & species	***Tachyoryctes splendens*** *others*
family	**Bathyergidae**
genera & species	***Cryptomys lugardi*** *blesmol* ***Heterocephalus glaber*** *naked mole-rat, others*

How to tell one end of a mole-rat from the other—by its two pairs of large front teeth!

Moloch

Moloch was a Canaanite god to whom children were sacrificed and was also one of Milton's devils. It is therefore not surprising that the name should have been given to this uncouth Australian relative of the pretty agama lizards (p. 59). The moloch is, however, much maligned by its name as it is a most inoffensive lizard, yet the Australian aborigines treat it with care as they believe it is harmful. The alternative names of the moloch are mountain devil or thorny devil. The latter is the most descriptive. The moloch is covered with thorn-like spikes over its head, body, tail and legs. They are triangular in section and as pointed as any rose thorn. The lizard's total length is 6 in. and the body is round, so a moloch looks like a walking horse chestnut burr.

Molochs are found in many parts of South and Western Australia and in Northern Territory.

Lizard with the hump

Molochs are prickly lizards that move slowly even when in a hurry. When frightened they tuck their heads between their front legs, presenting a thorny hump that stands on the back of the neck. It is, how-

△ *A walking thorn bush. The moloch, harmless in itself, is a mass of spikes to any aggressor.*
▽ *A visitation from the devil? An advancing moloch looms up over a fallen tree trunk.*

ever, difficult to see how this can enhance the general prickly reception a predator gets. Another suggested function for the hump was that it is a food store, but as it does not shrink when a moloch goes hungry this seems unlikely.

Molochs live in deserts and semi-desert regions and, as with so many other desert animals, although they are active by day their behaviour is adapted to avoid the worst of the sun's heat. They can also change colour but this ability is often exaggerated. When transferred from one background to another they change colour slowly, taking several minutes. Against a sandy background a moloch may be a dull light grey but against other backgrounds it is sometimes prettily coloured with orange, chestnut and black markings.

Painstaking ant eaters

A favoured method of feeding is for a moloch to sit by a trail of ants, flicking them up with its tongue as they run past. It has been estimated that they pick up 30−45 ants a minute, and that a moloch eats 1−5 thousand ants at one sitting, each one being picked up separately, so one meal takes a long time. Molochs eat little else but ants, taking only those without stings. Their jaws are weak but their teeth have complex serrated crowns which are well suited for crushing the hard outer skeletons of ants so their soft interiors can be digested.

Outsize eggs

The breeding habits of molochs are known from those that have laid eggs while being kept in captivity. Mating takes place in October and November, and eggs are laid in January. The maximum recorded clutch was 10 eggs, each about 1 in. long and $\frac{1}{2}$ in. wide—enormous eggs to be produced by a fairly small lizard like the moloch.

Before laying eggs the female moloch spends 2–3 days digging a nest in soft sandy soil. She does this according to a set pattern. Having started a hole and thrown out a small pile she removes the surplus by scraping earth backwards from the top, and gradually digs her way forwards into the hole. When she has reached the 'pit face' and dug out more soil she turns and goes back to the pile outside and starts again. In this way she continually throws the soil backwards, so the entrance and the tunnel are kept clear. The finished tunnel is 2 ft long, running downwards and ending 10 in. below the surface. It takes a long time to dig be-

cause the moloch often stops to rest. Having completed her task she lays her eggs at the bottom of the tunnel and fills it up again, leaving an air cavity around the eggs so the developing molochs can breathe. When the tunnel has been completely filled, the surface is levelled and swept so the entrance is concealed. The moloch then leaves the eggs to develop and the young to emerge on their own. Hatching takes place 10–12 weeks later and the young molochs, $2\frac{1}{4}$ in. long, dig their way out and disperse.

Dew-trap lizard?

It has been said that molochs can absorb water through their skin, because a drop of water placed on the back of one of them rapidly disappears. If this were so, the skin would be most unusual, because at the same time it must prevent water from leaking out, otherwise molochs would not be able to live in deserts. It is now known that the drop disappears because it spreads rapidly

by flowing along minute grooves in the skin. If the tip of the tail is dipped into water the whole of the skin becomes wet in a few seconds. When the water reaches its lips the moloch starts to sip it. No one has studied the use of this mechanism but it may be a way of collecting dew, forming on the moloch's body during the cold desert night.

class	**Reptilia**
order	**Squamata**
suborder	**Sauria**
family	**Agamidae**
genus & species	*Moloch horridus*

▽ *Lizard gourmet. A moloch spends a long time over its meal, picking each ant up singly. This meal may last it for several weeks.*
▽▽ *A groovy skin. Tiny grooves spread any moisture all over its body extremely quickly.*

John Warham

John Warham

Monarch

A wanderer among butterflies, the monarch is native to both North and South America, the three subspecies inhabiting the northern, central and southern parts respectively of the New World continents. It has spread westwards across the Pacific and is established in Hawaii, Tonga, Samoa and Tahiti and in Australia and New Zealand. The fact that the extension of its range has taken place within the last 120–130 years suggests that it has been helped by shipping, and the butterfly has sometimes been seen on ship's rigging. It now seems certain that it may occasionally fly the Atlantic, to Europe, helped on its way by persistent westerly winds.

Far-flying migrant

The monarch is the most celebrated of all migratory butterflies. It is the only one which makes a definite journey in one direction each year and returns along the same route the following year. In North America it is found in summer all over southern Canada and the northern United States. In autumn the monarch butterflies in the north gather in groups and begin to move south. As they go, the groups are joined by other monarchs, so they get larger and larger until there are thousands of butterflies on the move. Each night they settle on trees, moving on in the morning. When they get as far as Florida in the east, and southern California in the west they settle in great numbers on trees and pass the winter in a state of semi-hibernation, occasionally flying round on warm, sunny days. The inland streams fly into northern Mexico. The same trees are used as resting places year after year, and in some places, where the butterfly trees are regarded as a tourist attraction, the hibernating insects are protected by law from disturbance.

In spring the butterflies of both sexes fly northward, the females laying eggs as they go, but on this return migration they fly

Roy Pinney: Photo Library Inc

△ *Mass formation. Migrating monarchs gather in profusion before moving off in ones and twos.*
◁ ▽ *Close up of a scent patch on hind wing of male African monarch.*

▽ *Turning it on and brushing it off: the scent patches on the hind wings and the scent brushes on the tip of the abdomen of an African monarch butterfly,* **Danaus chrysippus.**

Anthony Bannister: NHPA

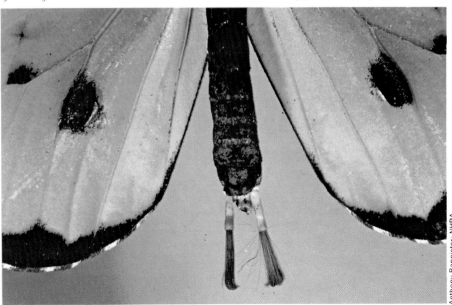

Anthony Bannister: NHPA

singly and in quite a dilatory manner, so their movement is far less easy to observe. During the summer two or three generations are passed through, and the butterflies of each generation continue to press northward. They do not reach Canada until June, and these butterflies are likely to be the grandchildren or great-grandchildren of those which left the previous autumn. The journey of a monarch from the extreme north of its range to the southern hibernation trees may be as much as 2 000 miles.

Unfit for consumption

Like other members of its family, the monarch is distasteful and poisonous to birds and other insectivorous animals; they leave monarch butterflies severely alone after one or two experiences of trying to eat them. The few monarchs that fall foul of predators are sacrificed for the sake of the many.

The larvae are conspicuously coloured with white, yellow and black stripes and have two pairs of black fleshy filaments on the fore and hind parts of the body, which make them even more unmistakable. They feed on various plants of the milkweed family (Asclepiadaceae), all of which are poisonous. They derive a two-fold benefit from this; firstly, the plants are avoided by grazing animals so the caterpillars do not run the risk of being accidentally eaten or otherwise destroyed, a fate that may overtake insects feeding on ordinary plants. Secondly the toxins of the plant are kept in the haemolymph or 'blood' of the insect, so it is poisonous in all stages. The rounded pupae, often brightly coloured with metallic gold markings, hang from a twig or some other support by a cord of silk from the tips of their tails.

The butterfly is strikingly marked and coloured on both the upper and under sides of the wings, so it is conspicuous both when flying and when at rest. This contrasts with what we find in most colourful butterflies, which are brightly patterned only on the upper side and this is concealed when they are resting. The monarch is a good example of warning colouration. Instead of concealing it, its colours make it conspicuous, so predators quickly learn to recognise and avoid it. The caterpillar's colouration protects it in the same way. The poison is a heart poison or 'cardenolide'. If retained in the predator's body it may be fatal, but it also acts on the stomach, so a bird swallowing the insect generally vomits it up after 10 – 15 minutes. In either case it is unlikely to make a meal of another one.

Tough monarch

The monarch and other Danaid butterflies have another interesting feature which has developed in association with this form of protection. They are extremely leathery and resistant to injury. Most butterflies can easily be killed by pinching the thorax, but it is almost impossible to kill a monarch in this way. Birds catch them, peck them and then, realising their mistake, release them, and the insect almost always flies away unharmed. An ordinary palatable butterfly, caught in this way, would be pecked to pieces and eaten, and any resistant properties would merely prolong its suffering.

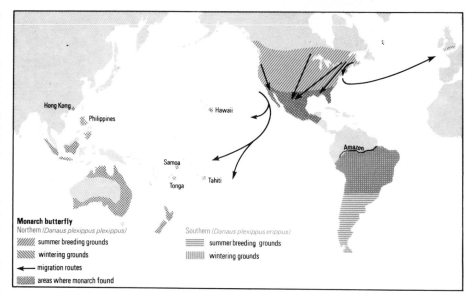

Monarch butterfly
Northern *(Danaus plexippus plexippus)*
///// summer breeding grounds
\\\\\ wintering grounds
→ migration routes
▨ areas where monarch found

Southern *(Danaus plexippus erippus)*
≡ summer breeding grounds
||||| wintering grounds

Deceptive viceroy

There is another American butterfly, the viceroy *Limenitis archippus*, which belongs to another family, the Nymphalidae, and is related to the European white admiral *L. camilla*. Although it is not distasteful to birds, it is coloured and patterned very like the monarch and is much the same size, so the two are difficult to tell apart except by close inspection. We know it belongs to the genus *Limenitis* by its anatomy and life history. There is no doubt that this butterfly gains protection from its resemblance to the poisonous monarch. In America captive jays, that have never before seen a butterfly, will readily kill and devour viceroys. If, however, they are experimentally offered monarch butterflies they may eat the first, and possibly a second, but soon learn to leave them alone. Moreover, they will then avoid any viceroys offered them.

Trans-Atlantic flights

Although monarch butterflies have appeared from time to time in Britain for many years, and the species is 'on the British list', it was thought until recently that they must cross the ocean on ships. In October 1968, however, a large number of monarchs was observed in southwestern England and at about the same time there was an unusual number of records of North American birds in Britain. Study of the meteorological conditions has shown that, just at the time the monarchs were swarming southward in the eastern United States, very strong westerly winds prevailed across the Atlantic, and it seems likely that both butterflies and birds were actually carried across. In 1933 there was a similar invasion which may well have been due to similar weather conditions. The absence of any suitable food plant makes it impossible, however, for the monarch butterfly to establish itself in Britain.

phylum	**Arthropoda**
class	**Insecta**
order	**Lepidoptera**
family	**Danaidae**
genus & species	**Danaus plexippus**, *others*

Eric Elms: NHPA

PH Ward

1 2

Anthony Bannister: NHPA

3 *(Opposite, top) This migratory butterfly has been found in many places at one time or another, as it is sometimes blown off course by strong winds when it is migrating. In most of the areas where it is found which have been marked on the map, it has become well established and breeds there, although in Britain it is just an irregular migrant, as in a few other places not marked.*

(Opposite, bottom) Keeping tags on their movements. Here tagged monarchs are just being released. The complete distribution of this butterfly is still unknown.

1 *The beginning of the saga. A monarch laying her eggs in the wild in Australia.*
2 *Beauty at an early age. A monarch larva.*
3 *An awkward time of life. The larva of a monarch **Danaus chrysippus** prior to pupation.*
4 *Hanging in the balance. The developing pupa of a monarch **Danaus plexippus**.*
5 *A monarch butterfly casts off its cloak of confinement as it emerges from the pupa.*
6 *A free butterfly **Danaus chrysippus** equips itself for its new life as it fits the two halves of its proboscis together.*

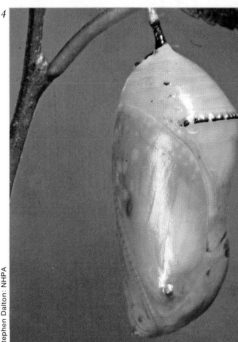

Stephen Dalton: NHPA

5 6

TR Priddle

Anthony Bannister: NHPA

Mongoose

A creature of many tastes, the mongoose is, to some people, a remorseless crusader against snakes; to others, just a sneak egg-thief and a destroyer of native wildlife. Mongooses live in southern and southeast Asia, Africa and Madagascar. There are 48 species, all much alike in shape, and most of them are not unlike in habits. The suricates of Africa are somewhat specialised mongooses which will be dealt with later. Mongooses are long-bodied with short legs, sharp muzzles and a tapering bushy tail. Their eyes are of moderate size and their ears small and almost hidden in their long, coarse fur. Most mongooses have 5 well clawed toes on each foot. The largest mongooses are 3½–4 ft long, of which just over half is head and body. The best known is the Egyptian mongoose, formerly called the ichneumon, which is 3½ ft long. It was venerated by the Ancient Egyptians and is still found from Egypt to the Cape as well as in southern Europe. The crab-eating mongoose of southeast Asia, Nepal to southern China, Burma and Malaya, is 4 ft long. The smallest is the dwarf mongoose of Africa south of the Sahara, 1½ ft long at the most, but usually smaller than this. Mongooses usually have a uniform grey or brown speckled coat or one peppered with light grey or white, except such species as the banded or zebra mongoose, which has more than a dozen dark transverse bands across the back, and the broad-striped mongoose of Madagascar which has half-a-dozen dark, conspicuous stripes along the back and flanks. In no mongoose are the underparts markedly lighter in colour than the back and in most the underparts and the legs are black.

Solitary or in colonies

All mongooses are alert and active, with very quick movements, and when they run along the ground the body and tail are in one line, so they seem almost to glide, especially when going through grass where the legs are completely hidden. They also spend long periods of time resting and relaxing and are especially fond of basking in the sun.

Most mongooses lead solitary lives, perhaps going about in pairs and usually then only when breeding. A few are gregarious, hunting in bands or living in colonies. The dwarf mongoose goes about in groups of up to a dozen, wandering around by day and feeding. At night they shelter in burrows, hollow logs, among buttress roots or in termite mounds. The cusimanse lives in groups of 10–24, the yellow mongoose lives in colonies of up to 50, in the abandoned burrows of other animals or in burrows they dig themselves. The various species of mongooses occupy a wide variety of habitats, from lowland to highland, forest to bush, savannah to semi-desert. The water mongoose is aquatic and the crab-eating mongoose gets much of its food in water.

△ *Directions from above. A half tame mongoose is gently but firmly guided by its Indian master.*
▽ *Underarm bowler. A dwarf mongoose aims an egg through its legs at the wall behind.*
All out! A triumphant but defiant look from the dwarf mongoose in its eggy glory ▽

J Allan Cash

Gerald Cubitt

Gerald Cubitt

△ *Rising tension as a cobra and mongoose confront each other on the battlefield.* △ *Death-lock as the mongoose leaps at lightning speed.*
▽ *Victorious mongoose devours its prey headfirst. Mongooses will often join forces to kill a snake, but are frequently successful in single combat. Snake venom has but a relatively mild effect on mongooses and their immunity increases every time they survive a bite.*

Most are ground-living but a few species are known to climb trees occasionally and the Madagascar ringtailed mongoose does so habitually. Most mongooses are active by day, the white-tailed mongoose being an exception, but a few seem to be active for parts of both day and night.

A crush on eggs

There are two particular foods for which mongooses are noted. One is snakes, the other is eggs. A trick used by some mongooses for breaking eggs is to throw them with the forepaws under the body, through the hindlegs onto a rock or wall. A few species will rise semi-erect, holding the egg in their forepaws, then crash it to the ground. Some mongooses will crack snails in the same way, and tame mongooses will take a billiards or pingpong ball and spend long periods of time simply crashing it against a wall. The main diet consists of almost any small animals, such as insects, lizards, birds and small mammals, to which some mongooses add frogs, fish and crabs. Several species eat fruit or a small amount of leaf. The dwarf mongoose drinks by dipping a forepaw into water and licking it. A tame dwarf mongoose will also eat soft foods such as egg custard with the same technique.

Information needed

Considering how numerous and well-known mongooses are and how often they have been kept as pets, surprisingly little

The map shows the distribution of three of the better known mongooses. Others include the marsh mongoose which lives throughout Africa south of the Sahara, the yellow mongoose which lives in the southern part of South West Africa and southern South Africa, and the kusimanse which lives in west Africa from Guinea to Ghana.

Mongoose
Egyptian *(Herpestes ichneumon)*
Dwarf *(Helogale parvula)*
Crab-eating *(Herpestes urva)*

is known of their breeding habits. It seems possible that they breed at almost any time of the year. Gestation lasts about 60 days and there are 2–4 babies in a litter. Even this is known for only a few species.

Enemies, too, are little known. Sometimes the remains of a small or medium-sized mongoose have been found under circumstances suggesting the animal was killed and eaten by an eagle. Although mongooses kill snakes, they sometimes die from eating them. Several have been found dead and post mortem examination has shown that they have eaten a snake whose fangs have punctured the wall of the stomach so that the poison has entered the bloodstream. This is surprising because they are known to have a high immunity to snake venom.

It was for this ability as a killer that the lesser Indian mongoose was taken to Hawaii and later to Jamaica, and from there to other West Indian islands, to kill rats and snakes. Wherever it has been introduced it has become a pest, a menace to the small native animals and to poultry.

Instinctive enemy

In addition to their immunity to the poison, mongooses have other ways of keeping out of trouble, as a tame one showed. Given a life-like rubber snake, it immediately attacked it. This dwarf mongoose was handreared as a baby, from before it had been weaned, so it had never before fought a snake. We can be reasonably sure that it had never even seen one. Certainly, since it was taken to England and kept as a pet for a year, it had had time to forget about snakes—if it had ever known one. Yet the moment the rubber snake was put on the ground near it, it leaped at the back of its head, sank its teeth into it and leaped away again, all at lightning speed. It attacked again and again, always in the same way and at the same lightning speed. A live snake could hardly have had a chance, either to dodge or to strike.

class	**Mammalia**
order	**Carnivora**
family	**Viverridae**
genera & species	*Atilax paludinosus* water mongoose
	Crossarchus obscurus cusimanse
	Cynictis penicillata yellow mongoose
	Galidia elegans ringtailed mongoose
	Galidictis striata broad-striped mongoose
	Helogale parvula dwarf mongoose
	Herpestes auropunctatus lesser Indian mongoose
	H. ichneumon Egyptian mongoose
	H. urva crab-eating mongoose
	Ichneumia albicauda white-tailed mongoose
	Mungos mungo banded mongoose others

◁ *Squatters asserting their position. Several dwarf mongooses take possession of a termitarium, their slender bodies enabling them to squeeze into fairly small holes.*

Jane Burton: Photo Res

Monitor lizard

The family of monitor lizards includes the largest lizards now in existence; several species reach 5 ft or more. The largest is the Komodo dragon (p. 1387) at 10 ft long, and the smallest is the 8-in. short-tailed monitor of Australia. The word monitor originally meant 'one who admonishes others about their conduct'. It was applied to these lizards because of an error in translation. The Arabic name for the lizards is **waran** which is very similar to the German **warnen** meaning 'warning'. As a result these lizards became known as 'warning lizards'. In Malaya and Australia monitors are called iguanas or 'goannas' although the iguanas form a separate family.

Monitors live in the warmer parts of the Old World, including the whole of Africa excluding Madagascar, Asia from Arabia to southern China down through southeast Asia to the East Indies and Australia. Unlike other lizard families where some species carry ornate crests, spines and frills, the monitors show very little variation in form. The snout, neck, body and tail are all long and slender and their eyes are prominent, making the lizard seem long, sinuous and alert. Monitors have several features in common with snakes although they are not so closely related to them as the earless monitors (p. 814). Both snakes and monitors have long, forked tongues and have lost the ability to shed their tails.

The largest monitor, excluding the Komodo dragon, is the two-banded or water monitor, at 8—9 ft. **Varanus giganteus**, the perenty, reaches 7 ft, and the Nile monitor 6 ft. The Nile monitor ranges throughout Africa from the Upper Nile south to the Cape, and the water monitor lives in Malaya.

The Nile monitor is brownish or greenish grey above with darker markings and yellowish spots which are lost with age. Underneath the skin is yellowish with dark bands. The Indian monitor is light to dark brown, sometimes with scattered light and dark scales on its back. Its underparts are dirty white with speckling. Young monitors, and adults just after they have shed their skins, are more brightly coloured. Young Indian monitors are orange to light brown with alternating yellow and black bands across the back.

Fast to escape, fierce at bay

Monitors are large lizards which can get quite nasty if brought to bay. At first they inflate their bodies and hiss. Then they attempt to deter any attack by violently lashing the tail like a whip. Finally they may attack by grabbing their adversary with their powerful jaws and clawing with their feet. Dogs often come off worst in such an encounter and a large monitor is a formidable opponent for a man. They may spend much of their time, however, basking in the sun and when alarmed they can run very rapidly with the tail held in a characteristic curve with the tip just off the ground. The Nile monitor can outpace a man, especially in thick cover. Except during cold seasons monitors are extremely active animals.

Each species of monitor has a fairly restricted habitat. The perenty, the largest of the Australian monitors, lives among rocky outcrops in the deserts. The desert monitor lives in similar habitats in North Africa and western Asia. Most monitors can climb trees but some forest-dwelling species are completely at home in trees. The Nile monitor and water monitor have flattened tails that are used for swimming and the water monitor has nostrils near the tip of its snout which allows it to breathe when almost completely submerged. When disturbed, monitors make for their natural home. Nile monitors usually bask on a branch or rock from which they can drop straight into the water. When surprised away from water they usually head for the bank but will sometimes hide in a tree or hole. The Indian monitor and the desert monitor also hide in holes in the ground while the forest-living Bornean rough-necked monitor takes refuge in trees.

Egg thieves

Like snakes, monitors swallow their prey whole rather than chew it as do iguanas and other lizards. A water monitor, for instance, has been known to swallow a 6in. turtle whole. Snakes and monitors have a strong bony roof to the mouth which protects the brain from being damaged by the passage of large mouthfuls.

Monitors are carnivorous, eating a wide variety of animals, as well as carrion. The Indian monitor eats palm squirrels, musk shrews, lizards, snakes and invertebrates and the water monitor feeds largely on fish. The Nile monitor eats mainly insects, snails, crabs and other invertebrates and also catches small birds and mammals. In South Africa, the Nile monitor digs rain frogs out of their burrows, ignoring the poisonous secretions that deter other predators. All monitors have a fondness for eggs and as a result are often persecuted by farmers for

KB Newman

△ *An inelegant sunbather,* **Varanus exanthematicus,** *sprawls on a cycad* **Encephalartos.** *Most monitor lizards are sun worshippers, spending much of their time basking on any bare surfaces.*

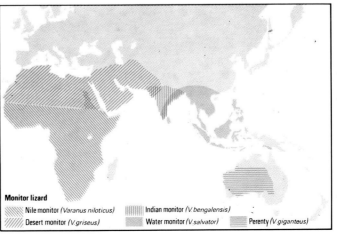

Monitor lizard

Nile monitor *(Varanus niloticus)*	Indian monitor *(V. bengalensis)*
Desert monitor *(V. griseus)*	Water monitor *(V. salvator)*
	Perenty *(V. giganteus)*

◁ *Each species of monitor has a fairly restricted habitat. The desert monitor and perenty, for example, live in arid country while the Nile monitor is never far from water.*

△ *Gas bag? A monitor lizard* **Varanus varius** *with its voluminous food pouch. It often swallows its prey whole.*

△ *A Nile monitor in its element. Although usually found in or near water, it can run rapidly and climb trees well.*

taking the eggs of poultry. The Nile monitor is said to steal the eggs from crocodiles' nests even though these are guarded by the mother.

Termite nest incubator

Monitors usually lay their eggs in nests in the ground or in hollow trees. The Indian monitor digs a hole about 1 ft deep with its forelegs then turns and backs in to lay its eggs. After laying it fills the nest hole by breaking down the sides with its snout and claws and tamping down the soil. In South Africa, Nile monitors lay their eggs in termite nests. They tear holes in the sides of the nests, lay their eggs in an enlarged chamber in the centre then leave the termites to repair the damage and so cover the eggs. Between 16 and 34 white leathery eggs are laid. The termites' nest provides heat for incubation and safety from enemies until the monitors hatch 5 months later. As the eggs hatch fluid escapes and softens the hard walls surrounding the nest. This enables the baby monitors to wriggle out of the eggs and then dig their way upwards. On reaching the surface they are reluctant to leave for several days, but eventually scuttle into nearby undergrowth.

Giant relatives

The monitors are among the oldest lizards, dating back some 130 million years. At this time other families of lizards that are now extinct were flourishing. Three families were closely related to the monitors: the aigialosaurs, dolichosaurs and mosasaurs. The latter two had taken to life in the sea, but the snakelike dolichosaurs did not last long. The mosasaurs were much more successful. The tail was flattened for swimming and the limbs formed broad paddles. The jaws were long and able to open very wide. Some mosasaurs reached a length of over 30 ft and preyed on fish and other reptiles that they seized in their large mouths with their sharp curved teeth.

The first fossil mosasaur was discovered in Maastricht in 1780 and consisted of a head over 3 ft long. The head attracted a considerable amount of attention and when the French attacked Maastricht in 1795 instructions were given to safeguard the house where it was thought that the fossil reposed. When the fortress fell it could not be found but a reward of 600 bottles of wine for its discovery soon led to 12 sol-

diers bearing the huge limestone block into the camp. Since that time mosasaur skeletons have been found all over the world. Many of them have fractured bones but it is a matter of debate whether these wounds were caused by fights with predators or in territorial battles.

class	**Reptilia**
order	**Squamata**
suborder	**Sauria**
family	**Varanidae**
genus & species	**Varanus bengalensis** *Indian monitor* **V. exanthematicus** *a South African monitor* **V. giganteus** *perenty* **V. gouldii** *Gould's monitor* **V. griseus** *desert monitor* **V. niloticus** *Nile monitor* **V. rudicollis** *rough-necked monitor* **V. salvator** *water monitor* **V. varius** *lace monitor* *others*

◁ *Top heavy? Close up of a head of a monitor lizard, alert and wary.*
▽ *Gould's goanna* **Varanus gouldii** *sticks its nose outside its front door.*

▽ *A bit of a sucker. A lace monitor* **Varanus varius** *tastes the air with its flickering forked tongue, like that of a snake.*

Monkfish

Though its shape is a clumsy compromise between a shark and a ray, the monkfish is a true shark, although, with a diet of shellfish, flatfish and worms it is hardly a dangerous one.

There are 11 species, all much alike. They can be up to 8 ft long. The head is broad and flattened, blunt and wide in front, with the mouth on the end, not on the undersurface as in sharks and rays. On top of the head, behind each eye, is a conspicuous crescent-shaped spiracle. The tail is more slender than that of a typical shark. There are two dorsal fins, well back on the tail. The gill-slits are crowded together in front of the pectoral fin, which is large and winglike, hence the alternative name of angelfish. Each pectoral fin is prolonged forward to form a 'shoulder' which is free of the head. The pelvic fins are also large and flattened and the shape of the monkfish when seen from above has earned the third name of fiddle

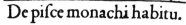

De pifce monachi habitu.

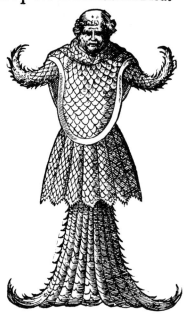

for centuries but their flesh was despised until fried fish shops became popular. They served other purposes. Their rough skin was used for polishing wood and ivory and for decorating trinket boxes, snuff boxes and sword hilts and for knife sheaths. Surprisingly it had a medicinal use: dried and powdered it was supposed to be a sovereign remedy for skin diseases.

What's in a name?

It is usually said that the name 'monkfish' is due to a fold of skin on the head that looks like a monk's cowl. This is hard to believe. The name came into use in the early years of the 17th century, not long after Rondelet published his hefty volume on fishes, *Histoire entiere des Poissons 1558,* in which was a somewhat grotesque drawing of a monk with scales. The picture, he said, had been given him by Margaret de Valois, Queen of Navarre, who had received it from a gentleman who gave a similar one to the Emperor Charles V. The gentleman declared he had seen the monster cast on the shore in Norway, during a violent storm. This could well be true, but in all probability

fish. The colour of the upper surface is yellowish to nearly black, usually grey or brown, blotched and with white lines and black and white spots. The undersurface is completely white.

Monkfishes are found in the subtropical and temperate seas on both sides of the Atlantic, in the Mediterranean, off South Africa, Australia, Japan and the Pacific coasts of North and South America. They sometimes wander into the shallower coastal waters during the summer.

Shark of the seabed
The only record of shark-like feeding in monkfish is of one that seized a cormorant by the wing and drowned it. They are normally bottom feeders. Their usual diet is flatfish, molluscs, crustaceans and worms,

but they also take other fishes, such as mullet. One monkfish was, however, found to have a mass of eelgrass in its stomach. Unlike the rays, which swim by flapping their large pectoral fins, monkfishes swim with a sculling action of the tail. They live mainly in shallow water, but some species live down to 4 200 ft and migrate into shallower water to drop their young. The young are born alive, in June and July, as many as 25 at a birth.

Dried skin for skin diseases
Although they are sharks, monkfishes are not normally dangerous. If landed and held by the tail they wriggle violently from side to side and snap their jaws, often giving any person holding one a demonstration of how well equipped it is to crush shellfish. Monkfishes have been well known to fishermen

Squatina squatina is the largest monkfish, reaching a maximum length of 8 ft and weighing 160 lb. Also known as angel fish and fiddle fish, this shark acquired the name monkfish after the publication of Rondelet's book in the late 16th century in which was a drawing of a monk with scales (above).

it was the distorted carcase of the monkfish as we know it which, with the aid of a little imagination, could easily assume a fanciful shape of some legendary monster.

class	**Chondrichthyes**
order	**Lamniformes**
family	**Squatinidae**
genus	*Squatina squatina*
& species	*S. californica, others*

Monk seal

Monk seals are the only true seals that live permanently in tropical waters, apart from the northern elephant seal (p. 847), yet they are most closely related to the Antarctic seals—the crabeater, leopard, Ross and Weddell seals. There are three species of monk seal, none of which is at all well-known as all are now very rare.

The Mediterranean monk seal was first recorded by Homer and Pliny, and Aristotle gives a detailed account of its anatomy. There are many myths about it and its skin was used as a protection against such varied evils as lightning, gout and insomnia. It is probably the largest monk seal, up to 9 ft 9 in. long. It looks rather like a young elephant seal and is chocolate brown on the back, shading to greyish on the belly. The Hawaiian and West Indian monk seals are 7–8 ft long; the Hawaiian is slaty grey above and silvery grey below, while the West Indian is greyish brown to yellowish white.

Unusual distribution

The three species of monk seal are widely separated. The West Indian monk seal is separated from the Mediterranean monk seal by the Atlantic Ocean, and from the Hawaiian monk seal by the Panama isthmus and most of the Pacific Ocean. Judith King, an authority on seals, has suggested that the monk seals spread from the west coast of Africa by the North Equatorial current that flows across to the Caribbean Sea. She also points out that it is not impossible for them to have crossed the Panama isthmus, long before the canal was built. Some seals, like the common seal (p. 636), travel up rivers, and others have travelled overland (crabeater seal, p. 688). The monk seal could have reached the Pacific in this way.

Although monk seals may once have travelled considerable distances, they lead sedentary lives nowadays. Most of their time is spent in shallow water, coming ashore during the day to bask. At times they must spend a considerable time at sea as individuals have been found with green algae growing in their fur, and this green colour disappears as soon as they come ashore to bask.

resting at times on their mothers' backs.

When their pups are born the mothers are extremely fat but they lose weight rapidly as they do not feed while they are with their pups. The pups grow very quickly at the expense of their mothers. Every time a pup gains 1 lb, its mother loses 2 lb. The pups double their weight in their first fortnight and at 5 weeks are four times their weight at birth and can hardly move. When their mothers abandon them, the pups survive on their stores of blubber while they learn how to catch their own food.

Extinct or hanging on?

It is always unwise to write off an animal as extinct as the stories of the dibbler and *Burramys* (p. 797) show. The possibility of the survival of the West Indian monk seal is extremely tantalising. This species was first recorded by Christopher Columbus in 1494 when his men killed eight 'sea wolves' on a rocky islet south of Haiti. Until the end of the 19th century they were abundant enough to be hunted for their blubber which was rendered into oil, but since then there have been only a few re-

◁ *Hawaiian monk seal and pup near weaning. It is abandoned by its mother at about 5 weeks, when 4 times its birth weight.*
▽ *Close-up of newly weaned Hawaiian monk seal. The pup will survive on its stores of blubber until it learns how to catch its own food.*

Karl W Kenyon

All the species have suffered severely from being hunted. The Mediterranean monk seal, once common in the Mediterranean and Black Seas, now occurs in only a few places such as the Island of Rhodes, Corsica and Cape Caliacra on the coast of Bulgaria. Outside the Mediterranean it is found along the West African coast as far south as Cape Blanc in Mauretania, where there is the greatest concentration, and there is a small colony on the Desertas Islands. The total world population is thought to be 500. The West Indian monk seal, once common in the Caribbean and the Gulf of Mexico, is possibly extinct. The Hawaiian or Laysan monk seal was once abundant but now numbers only about 1 500.

Fishy diet

The food of monk seals has not been studied in detail but in the Black Sea they are known to eat fish such as mackerel, plaice and flounder, especially the last. The West Indian and Hawaiian monk seals hunt around coral reefs. The Hawaiian monk seal is known to catch octopi, conger and moray eels as well as puffers, goatfish and various other flatfish and reef fish.

Rapid growth

The breeding habits of monk seals have been studied in Hawaii, where the pups are born in September and October. It is thought that this species mates in alternate years only. The females congregate on beaches to give birth but there is no harem as in elephant seals and grey seals. Pups are born with a black coat and swim at 4 days,

ports of seals in the West Indies. Some authorities have declared the species extinct. Two were seen a couple of miles south of Kingston, Jamaica in 1949 and since then lighthouse keepers and ships' crews have reported seeing seals or hearing their barks. It is quite possible that a few West Indian monk seals still live among the many rocks and islands in the West Indies.

class	**Mammalia**	
order	**Pinnipedia**	
family	**Phocidae**	
genus & species	***Monachus monachus*** *Mediterranean*	
	M. schauinslandi *Hawaiian*	
	M. tropicalis *West Indian*	

Moorhen

A member of the rail family, the moorhen is almost cosmopolitan and is a familiar bird on ornamental lakes in parks as well as in ponds and streams. The name moorhen is derived from 'mere', an Old English word for lake or pool, and the bird is rarely found far from water. The moorhen is up to 13 in. long with a plump body, small head and short tail. The legs are long and the long toes have no lobes or webs like those of other water birds but have flanges of skin which open when they swim. The plumage is brownish-black with white stripes on the flanks and white flashes on the underside of the tail. The bill is yellow with a red base and there is a red frontal shield over the forehead. The legs are green. Young moorhens are much browner and their frontal

In parks and near houses moorhens become quite tame and they can be seen walking over the grass with a high-stepping gait, running to water if disturbed.

Although they usually scutter across the water to safety, moorhens occasionally dive and swim underwater to cover. Young moorhens habitually dive while their parents flail over the surface, but as they grow older this habit is lost. When it surfaces after diving a moorhen may show only head and neck, like a diver (p. 774). After making sure that danger has passed, it rises until it is floating on top of the water. At one time there was some difference of opinion as to whether such a buoyant bird as the moorhen was able to submerge itself like a grebe or diver and assertions were made that moorhens held their bodies underwater by clinging to water plants. There is no proof that this is so and a moorhen has been seen walking on the gravel bed of a river. It could not possibly have been hanging on to the gravel to keep submerged.

△ *High and dry. A moorhen builds her nest on a platform to avoid it being waterlogged.*

roebild

◁ *Summer scene – a hungry moorhen chick stretches out greedily to get its food.*

shields are greenish-brown.

The moorhen inhabits the whole of Europe except northern Scandinavia and Finland, and western Asia and then northwards to Sakhalin, Japan, the Philippines and southeastwards to Java. It is also found in the Mariana Islands east of the Philippines. In Africa, the moorhen is found in most places outside the Sahara desert and in America it ranges from the Great Lakes of North America to northern Argentine.

High-stepping skulker

In the wild moorhens skulk amongst the reeds and low growth beside the banks and give themselves away by their loud 'kurruk' call. If one waits quietly they can be seen threading their way through the reeds and rushes with their distinctive white tail flashes flicking as they swim. When disturbed on open water they half fly, half run over the surface to the cover of the reeds.

Feeding on land and in water

Moorhens feed among water plants and also on land, searching for food on coarse pasturage. Duckweed, rushes and the fruits of water lilies are taken from the water and blackberries, elderberries and yew berries are taken on land. Moorhens usually eat fallen fruit but they can climb along quite slender branches to pluck fruit. About one quarter of the moorhen's food is insects and other small animals, including beetles, flies, caddis flies and snails.

Colourful chicks

Moorhens breed on small ponds, lakes, rivers and streams, providing the water does not flow too fast. Occasionally they nest some distance from water, sometimes in bushes or trees, where they use the abandoned nests of pigeons, rooks or magpies. Sometimes they maintain their territories all winter, otherwise territorial behaviour starts in February. Defence of the territory may be quite violent; rival moorhens rear

Moorhen *(G*

Jane Burton: Photo Res

▷ *Moorhen footprints in the mud. A clue even 'my dear Watson' could not miss!*

▽ *The moorhen is a cosmopolitan bird as the distribution map shows.*

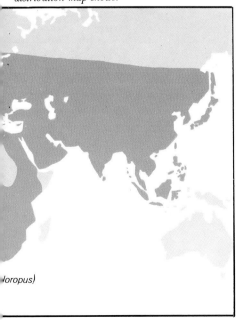

loropus)

up in the water and kick with their feet.

Before the proper nest is built, platforms are made among the reeds. The true nest is usually built amongst water plants, on the bank or a floating log. The foundations are made of lengths of reeds, flags and other plants, sometimes snipped off by the bill, level with the water. These are laid neatly, then other vegetation is piled on top. Sometimes flowers or pieces of paper are added as decoration and flags of iris drawn down to make a canopy.

The first eggs are usually laid in April, exceptionally as early as the beginning of March. The clutch is large, usually 5–11 eggs, whitish-grey to green with red-brown spots. The parents take turns in incubating them for 3 weeks. The newly-hatched chicks are covered with black down and have black legs. The frontal shield is bright red and the bill is orange with a yellow tip. There is a patch of blue skin around the eyes and pink skin shows through the sparse down of the back of the head. A few weeks later the

J-P Varin: Jacana

bill becomes green and does not attain the sealing-wax red of the adult until the following winter.

The chicks are unsteady on their legs at first and help themselves along with their stubby wings, each of which bears a claw like that of a hoatzin (p. 1211). They take to the water very soon after hatching and follow their parents, diving whenever the alarm is given. When they are 3 weeks old the chicks can feed themselves and they fly when 6–7 weeks old. They stay with their parents throughout the summer, while one or two more broods are raised.

Enemies all around

With 2 or 3 broods, each of 7 eggs on average, moorhens can rear potentially large numbers of young each year. Habits such as those of nesting in the reeds and swimming on open water, however, as well as the chicks' ability to dive seem to offer little protection. As the chicks grow up their numbers steadily decrease. Above the

water, herons are waiting to spear them; and underwater, pike engulf them. The predators do not always get their own way. An adult moorhen was seen attacking and successfully driving off a stoat that was swimming straight for its chicks.

Help from the family

Like coots (p. 654) young moorhens help their parents feed their younger brothers and sisters. Sometimes they have been seen to repair or add material to the nest when it has been in danger from flooding. RW Hayman records in *British Birds* how he watched a chick helping to add fresh material to a nest that had become detached from its moorings and had drifted into open water. An adult moorhen brought pieces of water plant and sticks and handed them to the chick. The chick worked the material into the nest while its parent went off for more.

This behaviour is reminiscent of the early stages of the development of social behaviour in insects, where a brood of young bees stay with their mother to help rear the next brood. It also shows how nest building is an instinctive skill. Most birds do not attempt to build nests until they are sexually mature, but here is a very young bird performing creditably. It is easy to see that this behaviour could be very important if there were sudden flooding and all 'hands' were needed to save the nest.

class	**Aves**
order	**Gruiformes**
family	**Rallidae**
genus & species	*Gallinula chloropus*

Moorish idol

This is one of the most striking of the small reef fishes. It has given inspiration to artists, designers and decorators and has been figured on wallpapers and fabrics. It is said to have been sold in the fish markets of Hawaii, and may still be sold there. It is mentioned and depicted in almost every book on fishes. So although practically nothing is known of its way of life or its life history, it deserves a place in an encyclopedia of animal life.

There are three species of moorish idol, the largest being up to 8 in. long but 4 in. is more usual. The body is strongly flattened from side to side, and when looked at from the side it is nearly circular. There is, however, a high dorsal fin in the mature fish and a triangular anal fin, making the outline almost diamond-shaped. In front the snout is drawn out and ends in a small mouth with small, very fine teeth in both upper and lower jaw. Two bony horns grow out over the eyes. The tail is short and it carries a tail fin that is almost triangular. The most striking feature is the colour pattern, the body being white and pale yellow with broad bands of brownish black running from top to bottom. The skin is shagreen-like, being covered with small sharp scales which make it feel almost like fine sandpaper.

Moorish idols spread halfway round the world in tropical seas; they are found from East Africa around the coasts of the Indian Ocean, the East Indies, Melanesia, Micronesia and Polynesia to the coasts of Japan and the various islands off the Pacific coast of Mexico.

Hiding among coral

These fishes are often seen in shallow inshore waters but their real home is on the coral heads in deeper waters, in coral lagoons, especially in the surge channels through coral reefs, and along the outer edges of the reefs. Although the outward appearance and colours, the shapes of their fins and other physical features have been repeatedly described in great detail in one scientific paper after another, nothing has been recorded of how they swim or what is their food. At best we can only deduce something of this. They probably swim by waving their tails, dorsal and anal fins, with the tail fin used as a rudder. This is how the butterfly fishes and marine angelfishes, to which they are distantly related, also swim. It is the way fishes swim that spend their time among irregular reefs or coral where quick movements and sharp turns are needed rather than swift forward movement.

Tweezer jaws

The narrow jaws of a moorish idol, with small teeth in front, act almost like tweezers. The snout is very like that of the butterfly fishes. From looking at moorish idols in aquaria, as well as guessing from the shape of the jaws and teeth, they probably feed on small crustaceans and other small invertebrates picked out with the 'tweezers' from small crevices.

Babies with streamers

The spawning times and mating behaviour are unknown. Young moorish idols are seldom seen, probably because of the difficulty of collecting them from among the coral heads. The few that have been caught show that the young fishes, up to $\frac{1}{2}$ in. long, have much the same shape as the adults but with long, low dorsal and anal fins and only one black band running from the top of the head, through the large eye to the throat, where the pelvic fins are situated, the small pectoral fins being just behind the eye. In the front of the dorsal fin are three spines. Two of these are very short but the third is long and thin, about $1\frac{1}{2}$ times as long as the body, streaming out behind. As the fish grows this gets shorter and finally the whole dorsal fin assumes the well-known triangular shape. Another feature of the young fish is that it has a knife-like spine behind each corner of the mouth. These drop off when it has grown to about 3 in. long.

Why moorish idol?

Not the least puzzling aspect of these fishes is their common name. 'Moorish' is usually associated with the western Mediterranean region, where the Moors are best remembered for the way they spread across North Africa and into Spain centuries ago. But moorish idols do not live there. Moreover, the Moors would have nothing to do with idols. A possible key to the origin of the name may lie in the association of the word 'Moorish', meaning Mohammedan, with the language of southern India on the coasts of which the fish does live. This word is now obsolete but it was once used by Englishmen living in that part of India. Perhaps it would have been better to have adopted the Hawaiian name for these fishes—*kihikihi*.

class	**Osteichthyes**
order	**Perciformes**
family	**Acanthuridae**
genus & species	***Zanclus canescens*** *others*

▽ *At home among the colourful corals—the moorish idol has a very striking colour pattern and an exquisitely shaped dorsal fin.*

Ben Cropp

James Simon: Photo Res

Moose

The largest living deer, moose are up to 9½ ft long and weigh up to 1 800 lb. They have long legs, standing 7¾ ft at the shoulder, with the rump being noticeably lower. The antlers, worn by males only, span up to 78 in. and have flattened surfaces with many tines or snags. The summer coat is greyish or reddish brown to black above, being lighter on the underparts and legs. The winter coat is greyer than the summer coat. The head is long with a broad overhanging upper lip, large ears and there is a tassel of hair-covered skin which hangs from the throat and is known as the 'bell'.

The moose lives in wooded areas of Alaska and Canada and along the region of the Rockies in northwest United States. Known as the elk in the Old World, the species lives in parts of Norway and Sweden and eastwards through European Russia and Siberia to Mongolia and Manchuria in northern China.

In summer moose spend a lot of time in water. A cow (above) wades out to look for food while a magnificent bull (below) stands at the water's edge.

Wilford Miller

Diets for winter and summer

Moose tend to be solitary but in winter a number of them may combine to form a 'yard'; an area of trampled snow in a sheltered spot surrounded by tall pines and with plenty of brushwood for feeding. They stay there until the food is used up and then move on to make another yard. Moose are most at home in well-watered woods and forest with willow and scrub, and ponds, lakes or marshes. Much of their time in summer is spent wading into lakes and rivers to feed on water-lilies and other water plants. By doing this they also escape to some extent the swarms of mosquitoes and flies. They will submerge completely to get at the roots and stems of water plants. In winter they use their great height to browse the shoots, leaves and branches of saplings. A moose will also straddle a sapling, bending it over to reach the tender shoots. Bark, particularly of poplar, is also eaten. The Algonquin Indian name *musee* (now moose) means wood-eater.

Growth of antlers

The bulls shed their antlers in December. In April to May new antlers begin to sprout and by August they are full grown and shedding their velvet. At first the exposed antlers are white but after a bull has rubbed them well on bushes and low branches they become polished and brown. A yearling bull has spikes 6—8 in. long; in a two-year-old they are forked, and by the time he has reached 3 years he has narrow hand-shaped antlers with 3—4 points.

Mother's apron strings

The breeding season, September to October, is marked by fighting between the bulls, who spar with each other with their antlers, normally doing little damage. They are polygamous, a mature bull mating with several cows.

The bulls bellow for females and, on hearing their answers, smash their way noisily through the thick brushwood to find them. Indian hunters used to call up the bull moose by using a birch bark trumpet to imitate the cow's call. The bull may stay with the chosen cow until her calf is 10 days old. The gestation is 240—270 days, after which 1—3 calves, normally 2, are born. The first time a cow gives birth she has a single calf, but after this twins seem to be the rule, with triplets rarely. The calf is a uniform reddish brown. The calves run with the mother at about 10 days old. For the first 3 days the calf is unable to walk much and the cow keeps close beside it, often squatting low or lying down for the calf to reach her udder, the calf calling to bring the mother to it when it is hungry. Calves remain with the mother for 2 years, by which time they are sexually mature. Moose are long-lived animals, sometimes reaching 20 years of age.

Formidable hoofs

The main enemies are bears and wolves, and to a lesser extent pumas, coyotes and wolverines, which prey on the young. An adult moose is a match for any of these, defending itself not with its antlers but by striking downwards with its large hoofs and then trampling on its opponent.

CJ Ott: Photo Res

Learning social graces

An outstanding feature of the life of any moose is that for part of the year it is solitary, and for part of the year it is gregarious. There is, therefore, a big change in its mentality; at one time it shuns its fellows, at another time it joins them. The secret is in its early training: the mother stays close to her calf when it is first born, as we have seen. As the weeks pass the area over which the mother may wander is gradually extended but all the time the calf keeps close to the mother, following at heel. If it does not, and shows signs of wandering away, she brings it back. Mother and child may be said to cling to each other, and the mother drives everything out of her territory, not only other moose but other deer as well as horses, and even people.

Left to itself, however, a moose calf will follow any deer, horse or man that comes by. So it seems that by nature a moose calf is sociable but by training it becomes anti-social. When the rutting season begins, however, the mother is less inclined to drive moose or anything else out of her territory, so very soon the calf has the company of its mother, a bull and one or more cows that the bull brings in. The calf has company now, and in fact plays a part in the courtship. If the bull is rude to it, its mother will leave him, taking her calf with her. Otherwise all is well and the calf stays with this small group.

The following spring, when the new calf or calves arrive, the older calf stays with the mother but now she makes it keep to the perimeter of her territory. At the next breeding season, in September or October, the calf, its brothers or sisters and its mother, will be joined by a bull and one or more cows, possibly with their calves. If the calf, now in its second year, is a male, it will have to keep its distance or the bull courting its mother will treat it as a rival. It will creep back to its mother any time the bull goes off to drive away a full-grown rival, but it will have to retreat again when he returns. If a calf is a female its own mother will treat it as a rival and drive it away, making it keep its distance whenever the bull, her mate, is near. So each calf learns to be solitary but also learns that it can enjoy the company of others at one time of the year, so it is readily able to live in a group for winter feeding, yet be solitary for the rest of the year apart from the rut.

class	**Mammalia**
order	**Artiodactyla**
family	**Cervidae**
genus & species	*Alces alces*

◁ *Evidence of moose—a barked aspen tree. During the winter when food is scarce the moose out of necessity eats the bark of trees, twigs and any small plants in the snow.*

▷ *Old velvet. The bull 'sheds its velvet' every autumn as the tissues and blood vessels which form the covering skin die off.*

△ A moray eel, one of the **Lycodontis** species, floats jauntily into view, flaunting its crest.

△ Stars and stripes? A starry moray eel **Echidna nebulosa** emerges from its hole in the coral.

Moray eel

One of the ogres of the deep according to early divers' stories, the moray eel has a far worse reputation than it deserves. There are 120 species, living in the warmer seas, especially around coral reefs. They range in size from 6 in. to 10 ft or more. The front part of the body and the head are bulkier than the rest of the body and the mouth is large. In most species the jaws are armed with slender sharp teeth, some of which are long, making the open mouth somewhat like that of a snake. The skin is thick and scaleless. The gill pouch has a small opening. There are no pectoral or pelvic fins but the dorsal fin starts just behind the head and is often high on the front third of the body. In some species, however, the dorsal and anal fins, and sometimes the tail fin as well, are very small and indistinct or even hidden under the skin, so increasing the snake-like appearance. Some moray eels are sombrely coloured but many have gaudy colours and patterns, and because of this they are sometimes called painted eels. Some morays are a uniform brown or olive, others are yellow with a brown network, or whitish with dark spots and blotches. The dragon moray of the Pacific is reddish brown with a bizarre pattern of white spots and bars

as well as dark blotches. The green moray is brown but its natural colour is usually obscured by a film of green alga covering the body. The zebra moray of the Indo-Pacific has a rich ochre body marked with 50−80 white rings.

Morays live in tropical and subtropical seas, down to 150 ft, and are rarely, if ever, seen in the open seas.

Ill-founded reputation

Morays have the reputation of being venomous, but so far there is no evidence to support this. They also have the reputation of being aggressive, of attacking bathers and divers, and people searching the reefs for lobsters, abalones and other shellfish, and even of taking a tenacious grip of a man's arm and holding him underwater till he drowns. The present-day evidence, especially from skin-diving biologists, is that morays usually try to avoid a man as anxiously as a man tries to avoid them. If cornered or speared, however, a moray will lunge and bite in a tremendous effort to escape. There are a number of authentic records of severe wounds sustained in encounters with morays. In attacking, the moray's actions seem to be like those of a venomous snake, rearing its head and the front part of the body and striking down. It will do this underwater and also when the front part of the body has been raised out of the water.

Moray eels spend the day in cavities and

crevices among rocky or coral reefs. They come out at night to feed and will come out by day if disturbed. Many morays feed on octopuses and GE and Nettie MacGinitie, well-known American marine biologists, have suggested that when a man puts his hand among rocks, searching for shellfish, his moving fingers look like the arms of an octopus, and this could be one reason why a moray lunges and bites. Others suggest that as morays eat shellfish of various kinds, when a human hand grasps one of these it catches the moray's attention so the eel lunges to grasp the food, and incidentally seizes the man's hand.

They must bolt their food

The food of many morays is made up of almost any animal, dead or alive, that it can swallow whole. This includes crustaceans, molluscs and fishes. The members of one genus *Echidna*, of which the zebra moray is one, have flattened grinding teeth for eating clams and sea-urchins. One reason why morays must take only food they can swallow quickly is that they need a continuous flow of water through the mouth for breathing. As a result, they seem to be panting, especially when actively exerting themselves. This is the strongest argument against the stories of morays holding men under water.

Irritable at breeding times

It seems possible that the people attacked by morays without having provoked them

▽ Moray eels do not always live up to their reputation, which condemns them as being extremely vicious and aggressive.

▽ A legless zebra of the deeps. This moray **Echidna zebra** is a deep ochre colour with 50−80 white rings around it.

△ *A painted eel? A Java moray eel daubed with spots. Many morays are strikingly coloured, hence their alternative name of painted eels.*

△ *The sensory pits which decorate the face of this moray eel contain sense cells which respond to vibrations in the water.*

may have been attacked during the eels' breeding season. When the *Kon Tiki* raft was wrecked on an atoll in the Pacific, its crew was chased from the lagoon by morays. The scientific members of the Royal Indian Marine Survey Ship *Investigator* had a similar experience on the Betrapar Atoll, in the Indian Ocean, in 1902. On that occasion it was noted that the eels were breeding. Apparently there is no breeding migration as in freshwater eels. The females lay large numbers of heavily-yolked eggs from which hatch ribbon-shaped leptocephalus larvae (see p. 824) a few inches long.

Flesh rarely poisonous

Morays are caught and eaten in many parts of the world. In the Mediterranean there is the time-honoured story of the Romans keeping them in specially built reservoirs near the sea, to serve at their banquets. The Romans are also reputed to have thrown the bodies of dead slaves, or even living slaves guilty of small misdemeanours, to be nibbled by the morays. These stories are, however, rather doubtful and the piscinas, as the reservoirs were called, probably held congers as well as, or instead of, morays. All this helps to build up and maintain the unjustifiably evil reputation of morays. Even the supposed venomous nature of these eels probably springs from the few occasions when people have died after eating their flesh. John E Randall, writing in *Sea Frontiers*, in 1961, tells how 57 people in the Mariana Islands sat down

to a meal of moray. They felt a scratchy sensation in the mouth and throat as they ate it and 20 minutes later some felt their lips and tongue growing numb. Ten minutes after that some were unable to speak and medical help was sought. Yet in spite of stomach pumps, by the next day their hands and feet were tingling, they vomited and found difficulty in breathing. Many of the men had convulsions, eleven became comatose and two died. When we remember that at one banquet given by Caesar 6 000 morays were eaten and that morays have been eaten in Mediterranean countries ever since, as they have in tropical and subtropical countries around the world, such misadventures must clearly be very rare.

Sea dragons

There is a suspicion that the sea serpent story is founded on the sightings of many commonplace objects seen at unusual angles at various times and. places. One could be the moray eel. For example, when disturbed they will sometimes swim at the surface with the forepart of the body and the head held high above the water. We do not know the greatest size to which they grow. Morays 7 – 8 ft long have been caught, but estimates of larger individuals are for 10 ft or more. A large moray with bizarre colours swimming like this, with the front end of the dorsal fin looking like a mane, would come very near the conventional picture of a sea serpent. The illusion would be helped

by the snakelike set of teeth seen as it opened and shut its mouth. Then there is a further adornment. Most bony fishes have 4 nostrils, in 2 pairs on the top of the head. In some morays, the nostrils have leaf-like flaps sticking up from them. In others they are tubular, with the rear pair of tubes standing well up from the top of the head.

Morays have a trick of throwing their bodies into a knot and letting this knot travel forwards to or backwards from the head. One used this trick to free itself from an octopus clinging round its head, slipping its body back through the loop to force the octopus' tentacles off its head. A moray hooked on a line will try to do the same to free itself and in doing so will sometimes climb the line, tail first. William M Stephens has recorded that off Palm Beach, Florida, three anglers jumped overboard as a large brown moray came writhing tail-first aboard their boat.

class	**Osteichthyes**
order	**Anguilliformes**
family	**Muraenidae**
genera & species	***Echidna zebra*** *zebra moray* ***Gymnothorax funebris*** *green moray* ***Muraena helena*** *Mediterranean moray* ***M. pardalis*** *Hawaiian dragon eel* *others*

▽ *The cleaners. The shrimps, which live with the moray eel in its hole, keep its skin clean by removing parasites.*

▽ *Panting moray. They often swim with their mouths open as they need a continuous flow of water through their mouths to breathe.*

Morpho

For every person who has seen living morpho butterflies there must be tens of thousands who have seen their wings in brooches, lockets, plaques, trays and objets d'art. It is the brilliant blues of the males that are particularly attractive.

*Morphoes are among the largest and most brilliant of butterflies and like many other brilliant butterflies they spend much of their time in or above the forest canopy. There are less than 50 species, all from tropical America, from central Mexico to southern Brazil. The largest of all, **Morpho hecuba**, is 7 in. across the wings. The upper surface of its wings is black with a broad band that shades from an intense orange brown to light yellow. **M. rhetenor** is a deep blue on the upper wing surface. **M. neoptolemus** is a rich blackish brown with broad bands of blue shot with a delicate violet. **M. sulkowskyi** has a more delicate colouring like mother-of-pearl, and the iridescent wings change with the angle of the light. These are the best known species and in all of them the upper surfaces of the wings are iridescent, while the undersurfaces of the wings are patterned with browns, greys, blacks and reds to harmonize with the background instead of standing out. In all species the females are less showy, their wings being cryptically patterned above and below; but their colours are also iridescent, and very beautiful.*

Beauty out of reach

Morpho butterflies fly through forest clearings flapping their wings lazily. Their wings are very large in proportion to their body and will carry them along at a fair speed. As each shaft of sunlight strikes the butterfly, its upper wing surfaces flash and then lose their brilliance as it goes into the shade. When it turns, the undersides of the wings show, hiding the iridescent colours so the butterfly temporarily disappears, only to flash into sight again as the sunlight catches its upper wing surfaces. This must be as disconcerting to an animal

△ *Captured beauty.* **Morpho hecuba hecuba** *above,* **Morpho hecuba cisseis** *below, sub-specific variation.*

pursuing it as it is to the butterfly collector trying to net it. What happens over the tree canopy we can only imagine. Morphoes have been seen by people in aeroplanes flying over the forests, flashing in the sun above the treetops. Probably if pursued there the butterflies melt into the foliage when danger threatens.

Butterfly farming

The caterpillars of the morpho butterflies are covered in hairs which are irritating to human skin. They live together in a web spun communally, but not in harmony because they are strongly cannibalistic. Those that survive eventually pupate together in the web. For gregarious caterpillars to be cannibalistic is most unusual. It suggests that the butterflies have few enemies at any stage in their life history and that this is the means whereby the population numbers are controlled. The communal webs doubtless protect the caterpillars from such enemies as there are. Caterpillars of one

species of morpho that are not protected by a web bunch together when they rest, looking exactly like an orchid.

Despite their irritating hairs large numbers of caterpillars are bred and reared commercially, the butterflies being farmed to supply the demands for the iridescent blue wings that are used in jewellery, pictures and other fancy articles. This is the only way to get large quantities of a butterfly that normally lives in such inaccessible places.

Fast colours

The trade in iridescent blue butterfly wings would never have flourished as it has done had the colours been likely to fade. The colours are 'fast' because they do not depend on pigments but are structural colours. In the early days of the microscope it was soon discovered that the wings of butterflies and moths were covered with scales. These are arranged like tiles on a roof. They form the powder that is left on one's fingers after handling one of these insects. As microscopes were improved and better lenses were made, it could be seen that the scales were not, like tiles or slates, flat pieces with smooth surfaces. They have a pattern of ridges, 35 000 to the inch, which give a rigidity to each scale, like the corrugations in roofing materials. The patterns of ridges on the scales are much more complicated than originally thought. Each ridge is made up of many thin layers slightly separated from each other. These layers are inclined at an acute angle to the plane of the scale. It is as if each ridge were a complicated set of Venetian blinds whose surfaces break up the white daylight into its various colours because they reflect back only light of a certain wavelength. It is this which, among other things, gives the iridescence, so the colours, being due to the physical structure of the ridges on the scales, are permanent.

phylum	**Arthropoda**
class	**Insecta**
order	**Lepidoptera**
family	**Morphoidae**
genus	*Morpho*

▽ *A lighter shade of pale. A series of morpho butterflies showing the changes in colour as the angle of the light alters.* **Morpho sulkowskyi sulkowskyi.**

△ *Resplendent in its lifesize beauty. Underside of* **Morpho hecuba hecuba,** *one of the largest of the morphos, with a wingspan of about 7 in.*

▽ *The overlapping scales on a morpho's wing, magnified 200 times.*   ▽ *A single corrugated scale with thousands of tiny ridges;* × *2 000.*

Mosquito

There are 3 000 species of mosquito living everywhere from the Tropics to Arctic latitudes, often in enormous numbers. While not all are troublesome to man, some species are notorious bloodsucking pests which transmit distressing diseases such as malaria, yellow fever, elephantiasis and filariasis.

Mosquitoes are slender-bodied insects, about $\frac{1}{4}$ in. long, with a single pair of narrow wings and long slender legs. In most of them the wing veins and the rear edge of each wing are decorated with small scales. The antennae are hairy in the female and copiously feathered in the male, except in members of the subfamily Dixinae. In most species the female has a sharp tubular proboscis adapted for piercing and sucking fluids, usually blood. Exceptions to this are again found in the Dixinae, which are also unusual in having transparent larvae, known quite appropriately as phantom larvae.

Basically 'mosquito' and 'gnat' have the same meaning, the first being Spanish, the second Old English. Today many people speak of small insects that 'bite' as mosquitoes, and similar but equally small, harmless insects, especially those seen in dancing swarms, as gnats. The confusion is the same among scientists—judging from popular books on the subject—who speak of **Culex pipiens,** the commonest mosquito in Europe, as the common gnat, but otherwise restrict the use of the word 'gnat' to the Dixinae.

There are two main groups of mosquitoes, the culicines and the anophelines, represented by the genera **Culex** and **Anopheles** respectively. The wings of culicines are transparent or slightly tinted, while the wings of anophelines are usually marked with dark and light spots or patches. Another difference is that the female culicine has a pair of very short palpi beside the long proboscis while the palpi of the anopheline female are as long as the proboscis. The best way of distinguishing the two is that when resting the culicine holds its body horizontal to the surface on which it is standing and the anopheline tilts its body upwards.

Eggs in any water

Mosquitoes lay their floating eggs in water, which may be fresh, brackish or salt, according to the species. With few exceptions each species chooses a particular kind of watery situation, which may be the margins of ponds or lakes, in ditches, seepages, waterfilled cartruts or hoofprints, polluted waters, water collected in holes in trees —usually at the top of a bole where branches fork—in aerial plants growing on trees or in pitcher plants. Water butts are often homes for mosquito larvae, but the eggs have even been seen in the water bowl put down for a pet dog. *Anopheles* lays single

John Clegg.

Heather Angel

Russ Kinne: Photo Res

Life afloat. Compact rafts of the eggs of the common gnat **Culex** beside a duckweed plant. Between them mosquitoes lay their eggs in almost every type of water, although the majority live in fresh water. Each individual species is, however, very limited in its habitat.
△ Living down under? A mosquito larva hangs from the surface in still water. It breathes through the respiratory tubes which reach up to the surface. The head, thorax and abdomen are clearly visible as are the mouth brushes which hang like a drooping moustache and sweep particles into the mouth. (7 × life size)
◁ Bulky but active. Mosquitoes spend only a few days in this pupal stage, whether the life cycle lasts a year or 10 days. (13 × life size)
▷△ Blood transfusion. An adult mosquito **Culex pipiens** sucks up blood through its tubular proboscis. Most of them are specialised in their habitat also at this stage, each species usually preferring a particular type of host. (14 × life size)
▷▽ Bloated with blood. A female mosquito **Culex pipiens** has just taken her last blood meal before she lays her eggs.

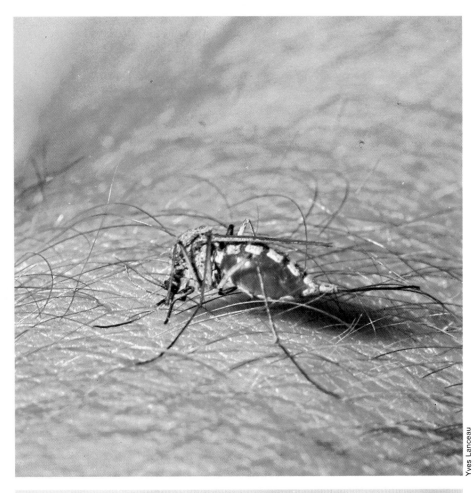

eggs, while *Culex* lays its eggs in compact masses or egg-rafts.

Each larva has a broad thorax in which all three segments are fused and an abdomen of 9 segments. The head bears the simple larval eyes and a pair of developing compound eyes. Brushes of bristles either side of it sweep fine particles of animal or plant food into the mouth, except in those larvae that extract dissolved food from the water, or prey on other insect larvae, usually other mosquito larvae. The thorax and abdomen are also decorated with long bristles. At the tip of the abdomen are four gills and a breathing siphon which can be pushed through the surface film of water to take in air. Some larvae feed on the bottom, others nearer the surface. They swim with twisting movements of the body, coiling and uncoiling spasmodically. The larvae rest just beneath the surface hanging down more or less vertically. At the slightest disturbance of the water they quickly swim down but, after a while, they must return to hang from the surface film in order to breathe.

Lively pupae

Larval life lasts about a week in most species, depending on temperature, but in those feeding on other insects it is prolonged, with usually only one generation a year. The pupae are active but do not feed, and the pupal life is short, at the most a few days. Pupae are typically bulky, having a large rounded head and thorax combined, with a pair of breeding siphons on top, and the abdomen more or less curled around it. In the last stages of development the pupa rises to the surface of the water, its hard outer skin splits, and the adult mosquito pulls itself out of the pupal husk and takes to the air.

The love call

Soon after leaving their pupal skins, the adults mate, after which the males die. The females must take a meal for their eggs to develop, either of blood or, In some species, nectar or sap. A few can manage on food stored during the larval stage. Some species of mosquitoes take the blood of mammals, others the blood of birds or even of amphibians. Sometimes a female will take another drink of blood after laying.

In the interval between leaving the pupal skin and mating the mosquitoes must rest. If a male takes to the wing too soon, his wings do not beat fast enough to proclaim him a male and other males will try to mate with him. He may lose some of his legs in the process. If a female takes off too soon, her wingbeats will be so slow the males will not recognize her until she has been in the air for a while and the pace of her wingbeats has quickened.

Surviving hard times

In temperate latitudes the females of some species pass the winter in sheltered places, such as caves, hollow trees or houses, especially in cellars. A few species lay their eggs in dry places which will be flooded in late winter or spring. These eggs can withstand dry and cold conditions, and in some instances will not hatch successfully without them. When the female of *Anopheles gambiae*,

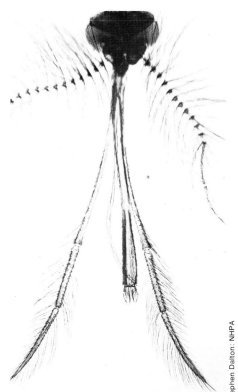

Stephen Dalton: NHPA

Arthur Christiansen

△ *The feathery antennae of the male (×60) are sensitive to sound vibrations, but only when the long hairs are erect. Some species keep them permanently erect so are always ready to mate while others erect them only at certain times of the day.* △ *Impending doom? A swarm of mosquitoes.*

a malaria-carrier, lives in desert areas, she gorges herself with blood and then shelters in huts, cracks in rocks or in rodent burrows, until the rains come. The dryness delays her egg-laying. Other species of desert mosquitoes lay thick-shelled eggs able to hatch even after 1–2 years, and in some cases up to 10 years later.

War on mosquitoes

Mosquitoes have many enemies. Airborne mosquitoes are eaten by birds such as swallows and flycatchers hunting on the wing. The larvae and pupae are eaten by small fish. The guppy (p. 1131), called the millions fish because of its large numbers in its native home, is used to control mosquito larvae and has been introduced into rivers in infested areas to keep down their numbers. Another control is to spray oil on waters of ponds and swamps. A small amount of oil will spread to cover a wide area with a thin film which prevents the larval mosquito from breathing at the surface of the water.

Homing on victims

When a female mosquito takes blood from a malaria patient, she passes the malaria parasite on to the next victim by injecting sporozoites in her saliva. It is the same with yellow fever, although a different species of mosquito is involved, and with elephantiasis, filariasis, and other mosquito-borne diseases. There are several defensive measures which can be taken to keep the mosquitoes away. This is usually achieved by netting or by deterrents, or by killing the larvae or the mosquitoes, or changing the habitat and reducing the number of people carrying the disease who act as a reservoir for further infection. The use of deterrents depends

on the behaviour of a mosquito in homing on its victim. An increase in the carbon dioxide in the air, as from human breathing, makes a female mosquito take off and fly upwind. As she draws near her victim the slight increase in temperature and humidity directs her more certainly towards her target until she can see where she needs to land. In these later stages the concentration of carbon dioxide is also greater, but certain chemicals (deterrents) will confuse her, put her off course and make her swerve away from her victim.

Mysterious outbreaks

The ague was once prevalent in Europe but was stamped out largely by the draining of the marshes. When soldiers serving in the tropics during the First World War returned home after contracting malaria, it was feared they might act as a reservoir for malaria (or ague) which would then become prevalent again in Britain. With the air traffic of today there is the fear that infected mosquitoes may be introduced into countries at present free from their diseases, and steps must be taken against them. Thirty years ago an African mosquito *Anopheles gambiae* found its way to Brazil, and 60 000 people died before the malaria was brought under control. In the Second World War, in Colombia, yellow fever suddenly struck villages where it had been unknown. In due course it was traced to the monkeys living high in trees which formed a reservoir, the carrier being a mosquito whose larvae lived in water in aerial plants growing in the tree tops. Woodmen felling some of these trees were attacked by mosquitoes from pupae in the aerial plants. The disease was then spread by a species of mosquito living at ground level.

Odd behaviour

Not all mosquitoes are troublesome; some are highly interesting. The larvae of *Mansonia* do not need to swim to the surface of the water to breathe. They have a saw-like apparatus for piercing water plant roots and drawing off the air contained in them. The females of another species *Leicesteria* lay their eggs onto their hindlegs which they then push through small holes in bamboo stems where water has collected. The eggs fall into the water and later hatch. A New Zealand mosquito *Opifex* has unusual mating behaviour. The males fly over water waiting for pupae to come to the surface to release the females within. They then mate with the females before they can get out. A mosquito *Harpagomyia* of Africa and southern Asia settles on tree trunks waiting for ants to pass. It then flies over to an ant, holds the ant's body with its front pair of legs and does not let the ant go until it has brought up a drop of food from its crop. The oddest story of all is, however, of a tropical American carrier of yellow fever which seems to prefer laying its eggs in water in flower vases, even those in hospital wards. In dealing with an outbreak of yellow fever in New Orleans it was found that the mosquitoes were breeding in the water in flower vases placed on the graves of the unfortunate yellow fever victims.

phylum	**Arthropoda**
class	**Insecta**
order	**Diptera**
family	**Culicidae**